Springer Tracts in Modern Physics
Volume 193

Starting with Volume 165, Springer Tracts in Modern Physics is part of the [SpringerLink] service. For all customers with standing orders for Springer Tracts in Modern Physics we offer the full text in electronic form via [SpringerLink] free of charge. Please contact your librarian who can receive a password for free access to the full articles by registration at:

www.springerlink.com

If you do not have a standing order you can nevertheless browse online through the table of contents of the volumes and the abstracts of each article and perform a full text search.

There you will also find more information about the series.

Springer
Berlin
Heidelberg
New York
Hong Kong
London
Milan
Paris
Tokyo

Springer Tracts in Modern Physics

Springer Tracts in Modern Physics provides comprehensive and critical reviews of topics of current interest in physics. The following fields are emphasized: elementary particle physics, solid-state physics, complex systems, and fundamental astrophysics.
Suitable reviews of other fields can also be accepted. The editors encourage prospective authors to correspond with them in advance of submitting an article. For reviews of topics belonging to the above mentioned fields, they should address the responsible editor, otherwise the managing editor.
See also www.springer.de

Stefan Keppeler

Spinning Particles – Semiclassics and Spectral Statistics

With 15 Figures

Springer

Dr. Stefan Keppeler
Matematisk Fysik, LTH
Lunds Universitet
Box 118
22100 Lund, Sweden
E-mail: stefan.keppeler@matfys.lth.se

Cataloging-in-Publication Data applied for

A catalog record for this book is available from the Library of Congress.

Bibliographic information published by Die Deutsche Bibliothek

Die Deutsche Bibliothek lists this publication in the Deutsche Nationalbibliografie; detailed bibliographic data is available in the Internet at dnb.ddb.de.

Physics and Astronomy Classification Scheme (PACS): 03.65.Sq, 05.45.Mt, 31.15.Gy

ISSN print edition: 0081-3869
ISSN electronic edition: 1615-0430
ISBN 3-540-01184-6 Springer-Verlag Berlin Heidelberg New York

Springer-Verlag Berlin Heidelberg New York
a member of BertelsmannSpringer Science+Business Media GmbH

www.springer.de

Typesetting: Camera-ready copy from the author using a Springer LaTeX macro package
Production: LE-TeX Jelonek, Schmidt & Vöckler GbR, Leipzig
Cover concept: eStudio Calamar Steinen
Cover production: *design & production* GmbH, Heidelberg

Printed on acid-free paper SPIN: 10894930 56/3141/YL 5 4 3 2 1 0

Preface

The book at hand is occupied with semiclassical methods for spinning particles and their applications in the theory of quantum chaos, in particular the description of spectral correlations.

The revival of semiclassical methods in physics started about 30 years ago with Gutzwiller's derivation of the semiclassical trace formula. The power of semiclassical methods, expressing quantum mechanical objects in terms of classical quantities, is particularly obvious in the field of quantum chaos where one searches for remnants of classically chaotic behaviour in the quantum world. The beauty of semiclassical techniques, however, is their ability to provide us with a clear and intuitive (classical) picture for complex quantum mechanical circumstances, a feature which makes them popular in many different fields of physics. In this book we illustrate that this remains true even when dealing with a genuinely quantum mechanical property like spin that has no a priori counterpart in classical mechanics.

The material presented in this book derives from – and is in large parts identical to – my PhD thesis *Spinning particles: Semiclassical quantisation and spectral statistics* (Abteilung Theoretische Physik, Universität Ulm, 2002) supervised by Dr. Jens Bolte. Him I would like to thank for his valuable advice and for sharing his deep insight into theoretical physics with me. I am also very grateful to Prof. Frank Steiner who has always accompanied my work with interest and helpful advice. He also suggested the publication of this book.

During the last years I have enjoyed many discussions with numerous people on various subjects related to those covered in this book. I would like to thank all of them and mention but a few (in alphabetical order): Dr. Daniel André, Dr. Arnd Bäcker, Prof. Sir Michael V. Berry, Rainer Glaser, Dr. Grischa Haag, Dr. Jonathan M. Harrison, Prof. Jon P. Keating, Dr. Jens Marklof, Dr. Francesco Mezzadri, Dr. Stéphane Nonnenmacher, Dr. Jonathan M. Robbins, Dr. Roman Schubert and Dr. Roland Winkler.

Furthermore, I would like to thank Ute Heuser and Jacqueline Lenz at Springer-Verlag for the editorial guidance.

The research this work is based on was carried out in the Abteilung Theoretische Physik at the Universität Ulm with a half-year interlude in the Department of Mathematics at the University of Bristol and at BRIMS,

HP-Laboratories, Bristol. I gratefully acknowledge financial support from the Deutsche Forschungsgemeinschaft and from the Deutscher Akademischer Austauschdienst.

I would like to thank the Mathematical Sciences Research Institute at Berkeley for the hospitality during the period when this work was completed.

Special thanks for constant support and encouragement go to my parents and in particular to my wife Dana and our daughter Annika.

Berkeley, February 2003 *Stefan Keppeler*

Contents

1 Introduction

This introduction is a perspective of topics related to those covered in the present work. More detailed references to the literature are given at the beginning of the different chapters.

What is semiclassical about quantum mechanics?

As the title indicates the present work is concerned with semiclassical methods for a genuinely quantum mechanical phenomenon, namely spin. First of all one might ask, why there should be any room for semiclassical investigations in the quantum mechanical world. Semiclassical methods are known for reproducing the "old" quantum theory of Bohr and Sommerfeld as the short wavelength limit of wave mechanics. Clarifying this relationship could be considered as a purely academic task, though very interesting in its own right since on the first sight the two theories bear little resemblance. Maybe more important is the characteristic of semiclassical methods of not only often yielding precise results but also always providing a clear physical picture. Moreover, the objects involved are sometimes (though not always) easier to handle than in the full quantum mechanical description. Probably the most important reason for dealing with semiclassical investigations in the field of quantum chaos stems from the fact that they allow us to give a quantum mechanical meaning to qualitative properties like integrability or chaoticity, which are primarily defined in the classical realm. In the present work we will see that even our understanding of an a priori purely quantum mechanical property like spin, which seems to have no immediate analogue in classical mechanics, can be improved by employing semiclassical methods.

How do you define this semiclassical limit?

In general the limit in which classical mechanics can be used to describe quantum mechanical properties is called the semiclassical limit. More precisely, in this work we will always refer to the semiclassical limit as the formal limit $\hbar \to 0$. For a particular physical system under consideration this has to be identified with some equivalent limit while $\hbar$ is kept fixed. For example, for integrable systems the limit $\hbar \to 0$ corresponds to the limit of large quantum numbers, which is Bohr's correspondence principle. For a particle moving freely in a domain $D \subset \mathbb{R}^d$ with some boundary conditions on ∂D the high energy limit $E \to \infty$ is equivalent to the semiclassical limit $\hbar \to 0$, whereas for the hydrogen atom $\hbar \to 0$ corresponds to $E \to 0$, i.e. the approach to the ionisation threshold.

What is a trace formula?

The most widely known semiclassical tool is the WKB-ansatz for the stationary Schrödinger equation, which leads to (a refined version of) the Bohr–Sommerfeld quantisation conditions. We will discuss the influence of spin in this setting in Chap. 5. The semiclassical tool, however, which we will mainly use throughout the work is that of semiclassical trace formulae, a field which was pioneered by Gutzwiller [1, 2]. A trace formula expresses the density of states of a quantum system in terms of a sum over periodic orbits of the corresponding classical system. Thus, in contrast to the explicit Bohr–Sommerfeld quantisation, it only gives indirect information on the eigenvalues. The advantage of trace formulae over Bohr–Sommerfeld quantisation is that the latter applies only for integrable systems, whereas trace formulae can be derived for any kind of classical dynamics, be they integrable, chaotic or mixed. Besides, there are applications, e.g. in mesoscopic physics, where one is not interested in single eigenvalues but rather in certain (weighted) sums over the whole spectrum, such that trace formulae prove even more handy in this context than explicit quantisation rules (see [3] for an overview).

Why is chaos relevant here?

A primary motivation for investigating semiclassical trace formulae comes from the field of quantum chaos, the search for fingerprints of classical chaos in quantum mechanics. A central observation is that the statistical distribution of energy eigenvalues depends on the properties of the corresponding classical dynamics [4]. Bohigas, Giannoni and Schmit (BGS) conjectured [5] that for classically chaotic systems the corresponding quantum spectral statistics can be described by random matrix theory, see e.g. [6, 7]. This hypothesis was verified numerically for many different systems. An explanation, which is built on the relation between a quantum system and its classical analogue provided by the Gutzwiller trace formula, was given by Hannay and Ozorio de Almeida [8] and Berry [9]. For examples which are known to violate the BGS-conjecture [10–12] it is again an analysis of trace formulae which shows why the arguments of [8, 9] do not apply in these situations. Another universality conjecture was put forward even before the BGS-conjecture by Berry and Tabor [13] who stated a relation between classical integrability and quantum spectral statistics which can be described by a Poisson process. Again the explanation [8, 9, 13] is based on semiclassical trace formulae and follows the same lines as in the chaotic case.

How do you include spin in this scheme?

Although spin is a genuinely quantum mechanical property it was demonstrated [14, 15] that the semiclassical programme which was initiated by Gutzwiller can also be carried out for multi-component wave equations describing spinning particles. It turns out that in leading semiclassical order the orbits in trace formulae are still solely determined by the translational dynamics. Spin enters through a weight factor which can be calculated from classical spin precession along the orbit. Classical equations for spin precession were already discussed in 1926 by Thomas [16, 17], but only a semiclassical treatment of the Pauli or Dirac equations clarifies the relevance of classical spin precession for quantum mechanics. The different levels on which translational and spin dynamics enter the formulae give rise to combined classical dynamics described by a skew product [18], see also [19]. This skew product defines an area-preserving but non-Hamiltonian flow whose characteristics need to be analysed in order to understand problems like the semiclassical quantisation of spinning particles or the role of spin in spectral statistics.

And where can I learn about this?

In Chap. 3 we review the derivation of semiclassical trace formulae with spin. Chapter 4 is dedicated to a detailed analysis of properties of the skew product of classical translational dynamics and spin precession, which allows us to proceed to the problem of torus quantisation for systems with spin in Chap. 5. In Chap. 6 we derive so-called sum rules for the skew product of translational and spin dynamics which are an important tool for the semiclassical analysis of spectral statistics presented in Chap. 7. There we treat the following three different situations. First the two cases in which the skew product is either ergodic or integrable and then an intermediate situation in which the classical translational dynamics is integrable but the skew product is not. A non-technical illustration of many concepts used in this work is given in Chap. 2 in terms of simple toy models.

References

1. M.C. Gutzwiller: J. Math. Phys. **12**, 343–358 (1971)
2. M.C. Gutzwiller: *Chaos in Classical and Quantum Mechanics* (Springer-Verlag, New York, 1990)
3. K. Richter: *Semiclassical Theory of Mesoscopic Quantum Systems*, no. 161 in Springer Tracts in Modern Physics (Springer-Verlag, Berlin Heidelberg New York, 2000)
4. S.W. McDonald, A.N. Kaufman: Phys. Rev. Lett. **42**, 1189–1191 (1979)
5. O. Bohigas, M.J. Giannoni, C. Schmit: Phys. Rev. Lett. **52**, 1–4 (1984)
6. M.L. Mehta: *Random Matrices*, 2nd edn. (Academic Press, San Diego, 1991)
7. T. Guhr, A. Müller-Groeling, H.A. Weidenmüller: Phys. Rep. **299**, 189–425 (1998)
8. J.H. Hannay, A.M. Ozorio de Almeida: J. Phys. A **17**, 3429–3440 (1984)
9. M.V. Berry: Proc. R. Soc. London Ser. A **400**, 229–251 (1985)
10. J.P. Keating: Nonlinearity **4**, 309–341 (1991)
11. E.B. Bogomolny, B. Georgeot, M.J. Giannoni, C. Schmit: Phys. Rev. Lett. **69**, 1477–1480 (1992)

12. J. Bolte, G. Steil, F. Steiner: Phys. Rev. Lett. **69**, 2188–2191 (1992)
13. M.V. Berry, M. Tabor: Proc. R. Soc. London Ser. A **356**, 375–394 (1977)
14. J. Bolte, S. Keppeler: Phys. Rev. Lett. **81**, 1987–1991 (1998)
15. J. Bolte, S. Keppeler: Ann. Phys. (NY) **274**, 125–162 (1999)
16. L.H. Thomas: Nature **117**, 514 (1926)
17. L.H. Thomas: Philos. Mag. **3**, 1–22 (1927)
18. J. Bolte, S. Keppeler: J. Phys. A **32**, 8863–8880 (1999)
19. J. Bolte, R. Glaser, S. Keppeler: Ann. Phys. (NY) **293**, 1–14 (2001)

2 Warming up: Oscillators

In order to become familiar with the language of trace formulae and in particular with the role spin plays in this context we will now describe some simple systems, namely the harmonic oscillator and generalisations, for which there exist exact trace formulae. These will be obtained in a direct way by applying the Poisson summation formula to the exact quantum mechanical density of states and only then be interpreted in terms of classical objects. The general derivation of semiclassical trace formulae certainly works the other way round. In a second step we will also demonstrate how semiclassical trace formulae can be applied in order to gain information on spectral statistics. In particular we introduce the two-point correlation function and its Fourier transform – the spectral form factor – and explain the idea of the diagonal approximation, which in these simple cases is exact.

2.1 Semiclassical Trace Formulae

One of the simplest models to be discussed in quantum mechanics is the harmonic oscillator, described by the Hamiltonian

$$\hat{H} = -\frac{\hbar^2}{2m}\frac{\mathrm{d}^2}{\mathrm{d}x^2} + \frac{m}{2}\omega^2 x^2 , \tag{2.1}$$

defined on a suitable domain in $L^2(\mathbb{R})$, where m and ω denote mass and frequency, respectively. Its spectrum is well known to be discrete with eigenvalues

$$E_n = \hbar\omega\left(n + \frac{1}{2}\right) , \quad n \geq 0 , \tag{2.2}$$

see e.g. [1], from which we can define the spectral density or density of states

$$d(E) = \sum_{n=0}^{\infty} \delta(E - E_n) . \tag{2.3}$$

Applying the Poisson summation formula yields ($E > \hbar\omega/2$)

$$
\begin{aligned}
d(E) &= \sum_{k\in\mathbb{Z}} \int_0^\infty \delta(E - \hbar\omega(n+\tfrac{1}{2}))\, \mathrm{e}^{2\pi i k n}\, \mathrm{d}n \\
&= \frac{1}{\hbar\omega} + \sum_{k\in\mathbb{Z}\setminus\{0\}} \frac{1}{\hbar\omega} \exp\left(\frac{\mathrm{i}}{\hbar} k \frac{2\pi E}{\omega} - \mathrm{i}k\pi\right) ,
\end{aligned}
\tag{2.4}
$$

where we have separated the average density $\bar{d} := 1/(\hbar\omega)$ from the oscillating terms.

The single terms in (2.4) can now be related to the corresponding classical system which is described by the classical Hamiltonian of the harmonic oscillator with frequency ω,

$$
H(p,q) = \frac{p^2}{2m} + \frac{m}{2}\omega^2 x^2 . \tag{2.5}
$$

The solutions of Hamilton's equations of motion

$$
\dot{x} = \frac{\partial H}{\partial p} = \frac{p}{m} , \quad \dot{p} = -\frac{\partial H}{\partial x} = -m\omega^2 x , \tag{2.6}
$$

with initial conditions $x(0) = x_0$ and $p(0) = p_0$ are oscillations with period $T = 2\pi/\omega$,

$$
\begin{aligned}
p(t) &= p_0 \cos(\omega t) - x_0 m\omega \sin(\omega t) \\
x(t) &= x_0 \cos(\omega t) + \frac{p_0}{m\omega} \sin(\omega t) .
\end{aligned}
\tag{2.7}
$$

The mean density of states is obtained by dividing the area $|\Omega_E|$ of the region in phase space accessible for the classical system at energy E, i.e. the energy shell

$$
\Omega_E := \{(p,x) \mid H(p,x) = E\} , \tag{2.8}
$$

by the area $2\pi\hbar$ of a Planck–cell. With (2.5) one immediately finds

$$
|\Omega_E| = \int_{\mathbb{R}} \int_{\mathbb{R}} \delta\left(\frac{p^2}{2m} + \frac{m}{2}\omega^2 x^2 - E\right) \mathrm{d}p\, \mathrm{d}x = \frac{2\pi}{\omega} \tag{2.9}
$$

and thus $\bar{d} = |\Omega_E|/(2\pi\hbar) = 1/(\hbar\omega)$. The oscillating terms, on the other hand, are related to the periodic orbits of the classical dynamics. From (2.7) one easily calculates the action

$$
S(E) = \oint p\, \mathrm{d}x , \tag{2.10}
$$

which is given by the area of the ellipse

$$
H(p,x) = E \quad \Leftrightarrow \quad \left(\frac{p}{\sqrt{2mE}}\right)^2 + \left(\frac{x}{\sqrt{\frac{2E}{m\omega^2}}}\right)^2 = 1 \tag{2.11}
$$

with semi-axes $\sqrt{2mE}$ and $\sqrt{2E/(m\omega^2)}$, i.e

$$S(E) = \frac{2\pi E}{\omega} . \tag{2.12}$$

Therefore, the terms in (2.4) which rapidly oscillate in the semiclassical limit $\hbar \to 0$ or, equivalently, in the high energy limit $E \to \infty$ can be written as $\exp((\mathrm{i}/\hbar)kS)$, thus being associated with k-fold repetitions of the (periodic) orbit, formally including negative ones.

The additional phase $-\mathrm{i}k\pi$ can also be given a meaning in terms of the classical periodic orbits. It counts the number of conjugate points along these orbits, i.e. the number of points where two trajectories which were started at the same initial position x with infinitesimally different momenta again become infinitesimally close in phase space. For the harmonic oscillator these conjugate points are focal points since actually all trajectories started at a given point x with arbitrary momentum p again meet at $-x$ after half a period $T/2 = \pi/\omega$ has elapsed. Thus, the number of conjugate (here focal) points encountered along a periodic orbit from time 0 up to the period $T = 2\pi/\omega$ is $\mu = 2$. This number μ is known as the Maslov index of the orbit. It appears in the trace formula as the phase factor $\exp(-\mathrm{i}\pi k\mu/2)$ which here is given by $\exp(-\mathrm{i}k\pi)$.

Finally, we can rewrite the quantum mechanical spectral density (2.4) as

$$d(E) = \frac{|\Omega_E|}{2\pi\hbar} + \sum_{k\in\mathbb{Z}\setminus\{0\}} \mathcal{A}_k \exp\left(\frac{\mathrm{i}}{\hbar}kS - \mathrm{i}\frac{\pi}{2}k\mu\right) , \tag{2.13}$$

which is composed of a mean part involving the volume of the energy shell Ω_E of the corresponding classical system and a sum over k-fold repetitions of the periodic orbit with action S and Maslov index μ, weighted by some amplitude $\mathcal{A}_k$.

It was found by Gutzwiller [2, 3] that this structure is true in general if in (2.13) the equality is replaced by an asymptotic relation in the semiclassical limit $\hbar \to 0$. Aside from giving explicit formulae for the amplitude $\mathcal{A}_k$ for some special integrable systems [4] he derived the general form [2] of the amplitude in the case of isolated and unstable orbits. The general formula in the integrable case was later given by Berry and Tabor [5, 6]. The case of the harmonic oscillator, where

$$\mathcal{A}_k = \frac{T}{2\pi\hbar} , \tag{2.14}$$

is a special case of both these formulae, since any autonomous one dimensional system is integrable and its periodic orbits for fixed energy E are isolated.

We will now investigate how spin changes this picture. To this end consider the simplest case of a non-relativistic neutral particle whose translational dynamics are given by a harmonic potential and which has a spin 1/2 coupling to a constant external magnetic field $\boldsymbol{B}$ through

$$-\frac{g}{2}\mu_B \boldsymbol{\sigma B} \, . \tag{2.15}$$

Here $\mu_B = e\hbar/(2mc)$ is Bohr's magneton, g denotes the spin g-factor of the particle and $\boldsymbol{\sigma}$ is the vector of Pauli spin matrices,

$$\sigma_x = \begin{pmatrix} 0 & 1 \\ 1 & 0 \end{pmatrix} , \quad \sigma_y = \begin{pmatrix} 0 & -\mathrm{i} \\ \mathrm{i} & 0 \end{pmatrix} \quad \text{and} \quad \sigma_z = \begin{pmatrix} 1 & 0 \\ 0 & -1 \end{pmatrix} . \tag{2.16}$$

Thus, we have to consider the Pauli Hamiltonian

$$\hat{H} = -\frac{\hbar^2}{2m}\frac{\mathrm{d}^2}{\mathrm{d}x^2} + \frac{m}{2}\omega^2 x^2 - \frac{g}{2}\mu_B \boldsymbol{\sigma B} \, , \tag{2.17}$$

whose eigenvalues are readily calculated to be

$$E_n^{\pm} = \hbar\omega\left(n + \frac{1}{2}\right) \pm \frac{g}{2}\mu_B |\boldsymbol{B}| \, . \tag{2.18}$$

Again using the Poisson summation formula as in (2.4) the spectral density reads ($E > \hbar\omega/2$)

$$\begin{aligned} d(E) &= \sum_{n=0}^{\infty} \left[\delta(E - E_n^+) + \delta(E - E_n^-)\right] \\ &= \frac{2}{\hbar\omega} + \sum_{k \in \mathbb{Z}\setminus\{0\}} \frac{2\cos(\frac{k\pi}{\omega} g \frac{\mu_B}{\hbar}|\boldsymbol{B}|)}{\hbar\omega} \exp\left(\frac{\mathrm{i}}{\hbar}k\frac{2\pi E}{\omega} - \mathrm{i}k\pi\right) . \end{aligned} \tag{2.19}$$

Compared to (2.4) the only differences are a factor of two for the mean density, which stems from the fact that due to spin there are on average twice as many eigenvalues, and a weight factor $2\cos(\frac{k\pi}{\omega} g \frac{\mu_B}{\hbar}|\boldsymbol{B}|)$ for the contribution of each periodic orbit. The latter can be interpreted in terms of classical spin precession along the orbit. According to the non-relativistic version of Thomas precession [7, 8] a classical spin vector $\boldsymbol{s}$ fulfills

$$\dot{\boldsymbol{s}} = -g\frac{\mu_B}{\hbar} \boldsymbol{B} \times \boldsymbol{s} \, . \tag{2.20}$$

After one period $T = 2\pi/\omega$ of the oscillation the spin vector has been rotated about an axis parallel to $\boldsymbol{B}$ by an angle $\alpha = \frac{2\pi}{\omega} g \frac{\mu_B}{\hbar}|\boldsymbol{B}|$. It is twice the cosine of half this angle (multiplied by the number k of repetitions) which appears in the trace formula. Thus, in a similar way to the translational dynamics, where only periodic orbits contribute to the spectral density, we find a maximal contribution if the spin vector is rotated by a multiple of 2π, whereas the weight factor is zero for angles $\alpha = \pi \mod 2\pi$. Hence with the notation of (2.13) we can write

$$d(E) = \frac{2|\Omega_E|}{2\pi\hbar} + \sum_{k \in \mathbb{Z}\setminus\{0\}} 2\cos\left(k\frac{\alpha}{2}\right) \mathcal{A}_k \exp\left(\frac{\mathrm{i}}{\hbar}kS - \mathrm{i}\frac{\pi}{2}k\mu\right) , \tag{2.21}$$

which is also the general structure of the trace formula for the Pauli equation [9, 10]. An arbitrary coupling of translational and spin degrees of freedom leads to a spin precession

$$\dot{\boldsymbol{s}} = \boldsymbol{\mathcal{B}} \times \boldsymbol{s} \tag{2.22}$$

instead of (2.20) with an effective field $\boldsymbol{\mathcal{B}}$ including, e.g., the contribution of spin–orbit coupling and external fields as felt by the spin along the orbit.

Notice that in the case of an arbitrary effective field – supposing that we have solved (2.22) and found the angle α for spin precession from time 0 to T – we can obtain a general semiclassical quantisation condition for one-dimensional systems with spin by now applying the Poisson summation formula to (2.21), yielding ($E > \hbar\omega/2$)

$$\begin{aligned} d(E) &\sim \sum_{n\in\mathbb{Z}} \int_{\mathbb{R}} 2\cos\left(k\frac{\alpha}{2}\right) \frac{T}{2\pi\hbar} \exp\left(\frac{\mathrm{i}}{\hbar}kS - \mathrm{i}\frac{\pi}{2}k\mu - 2\pi\mathrm{i}kn\right) \mathrm{d}k \\ &= \frac{T}{2\pi\hbar} \sum_{n\in\mathbb{Z}} \left[\delta\left(\frac{S}{2\pi\hbar} - n - \frac{\mu}{4} - \frac{\alpha}{4\pi}\right) + \delta\left(\frac{S}{2\pi\hbar} - n - \frac{\mu}{4} + \frac{\alpha}{4\pi}\right)\right] . \end{aligned} \tag{2.23}$$

Here we directly read off the modified quantisation condition

$$S = \oint p\,\mathrm{d}x = 2\pi\hbar\left(n + \frac{\mu}{4} \pm \frac{\alpha}{4\pi}\right) , \tag{2.24}$$

which includes the spin rotation angle α.

In Fig. 2.1 we show the density of states for the harmonic oscillator in a constant magnetic field with

$$\varepsilon := \frac{g}{\hbar\omega}\mu_B|\boldsymbol{B}| = 0.2 \tag{2.25}$$

and $\hbar\omega = 1$, calculated from the first 50 orbits, i.e. we cut off the sum in (2.21) at $|k| = 50$. We clearly observe the formation of peaks at $E_n^{\pm} = n + (1 \pm \varepsilon)/2$ as expected.

Since the trace formula for the spectral density $d(E)$ is to be understood in a distributional sense we can only qualitatively observe convergence to the expected result in Fig. 2.1. In order to get a more quantitative picture one can plot the spectral staircase, the integrated density of states

$$N(E) = \int_0^E d(E')\,\mathrm{d}E' , \tag{2.26}$$

which has steps at the position of each energy eigenvalue, the step size being given by its multiplicity. From (2.21) we obtain (recall that for our toy model neither $|\Omega_E|$ nor α nor $\mathcal{A}_k$ depend on E)

$$N(E) = \frac{2|\Omega_E|}{2\pi\hbar}E + \sum_{k\in\mathbb{Z}\setminus\{0\}} 2\cos\left(k\frac{\alpha}{2}\right) \frac{\hbar\mathcal{A}_k}{\mathrm{i}kT} \exp\left(\frac{\mathrm{i}}{\hbar}kS - \mathrm{i}\frac{\pi}{2}k\mu\right) . \tag{2.27}$$

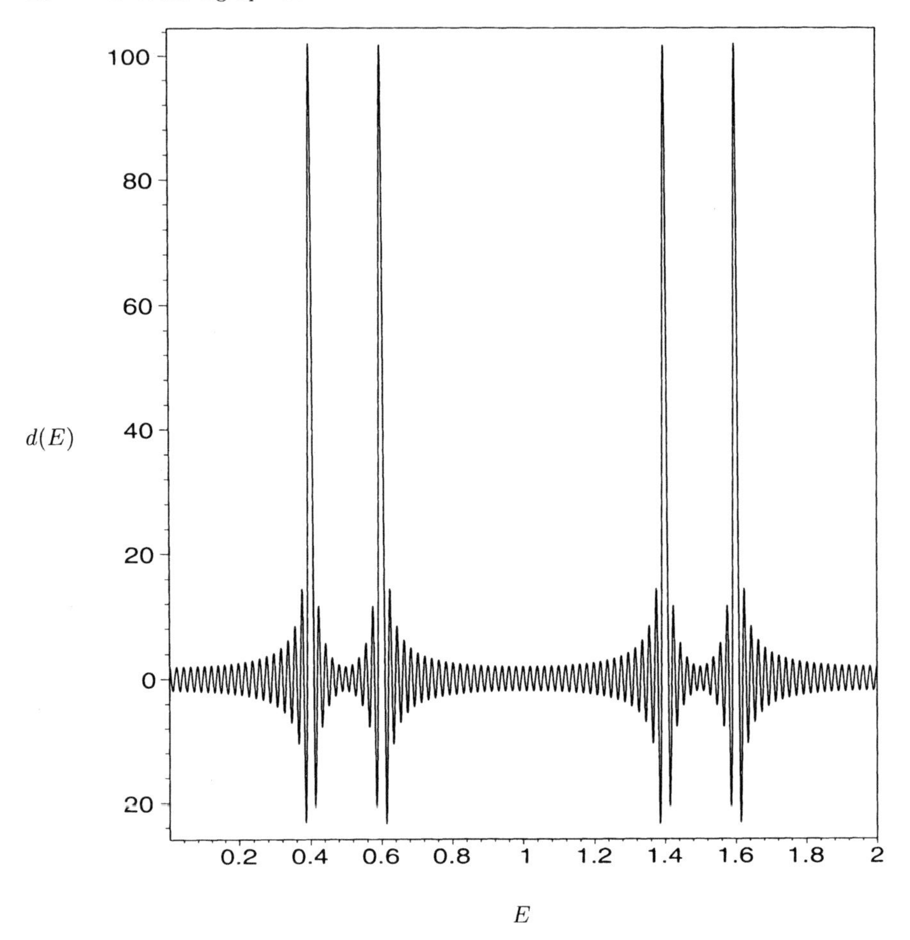

Fig. 2.1. Trace formula (2.21) for the spectral density of a harmonic oscillator with spin, with parameters $\varepsilon = 0.2$ and $|k| \leq 50$.

In Fig. 2.2 we show the spectral staircase for the harmonic oscillator in a constant magnetic field with $\varepsilon = 0.2$ calculated from the first 5 or 50 orbits, respectively. We observe convergence to the exact result which has steps of size one at $E_m^{\pm} = n + (1 \pm \varepsilon)/2$.

Having discussed the influence of spin in a non-relativistic context we also want to give an example of a trace formula for the Dirac equation. In one dimension the Dirac algebra can be realised by 2×2-matrices, i.e. the free Dirac equation reads

$$\mathrm{i}\hbar \frac{\partial}{\partial t} \Psi = \left(c\sigma_x \frac{\hbar}{\mathrm{i}} \frac{\partial}{\partial x} + \sigma_z mc^2 \right) \Psi \tag{2.28}$$

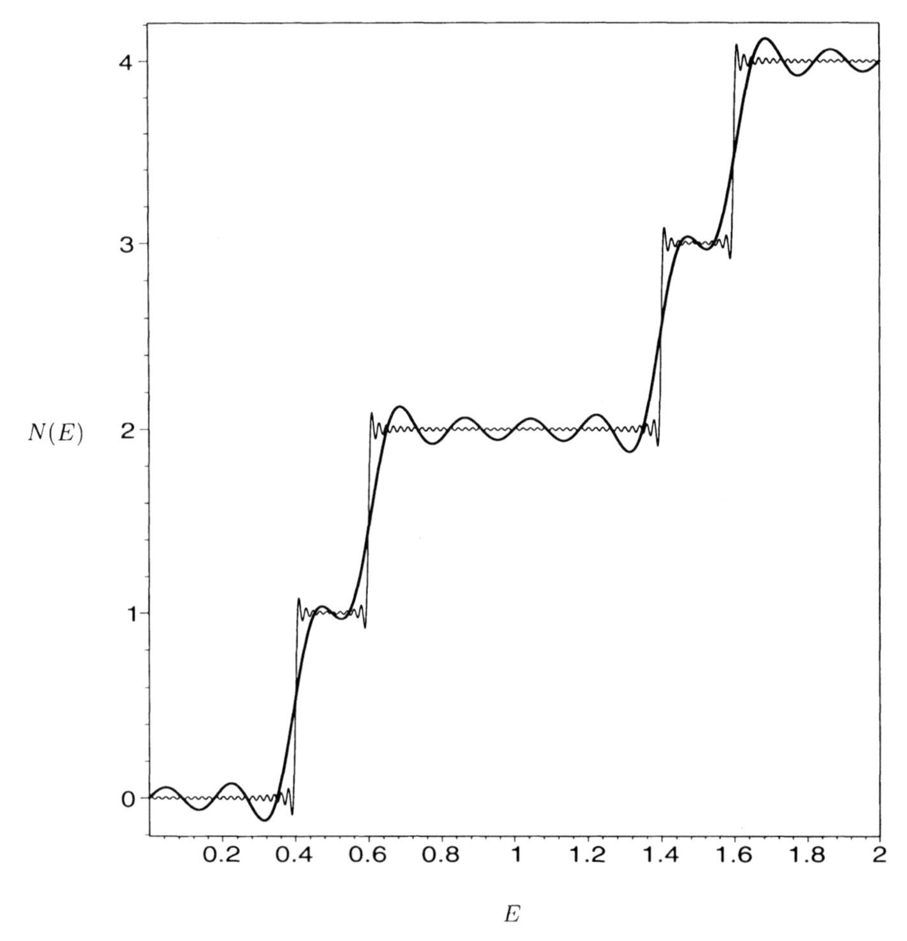

Fig. 2.2. Trace formula (2.27) for the integrated spectral density of a harmonic oscillator with spin, with parameters $\varepsilon = 0.2$ and $|k| \leq 50$ (*thin line*) and $|k| \leq 5$ (*thick line*), respectively.

for a two-spinor $\Psi \in L^2(\mathbb{R}) \otimes \mathbb{C}^2$. The particular model we want to discuss is a one-dimensional version of the so-called Dirac oscillator [11], which is obtained from (2.28) by replacing the momentum operator $\frac{\hbar}{\mathrm{i}}\frac{\partial}{\partial x}$ by $\frac{\hbar}{\mathrm{i}}\frac{\partial}{\partial x} - \mathrm{i}\sigma_z m\omega$. This model is called the Dirac oscillator, because in the non-relativistic limit $c \to \infty$ it reduces to the harmonic oscillator (2.1), see e.g. [11, 12]. If in the time independent Dirac equation $\hat{H}\Psi = E\Psi$ with the matrix valued Hamiltonian

$$\hat{H} = c\sigma_x \left(\frac{\hbar}{\mathrm{i}} \frac{\partial}{\partial x} - \mathrm{i}\sigma_z m\omega x \right) + \sigma_z mc^2 = \begin{pmatrix} mc^2 & c\frac{\hbar}{\mathrm{i}}\frac{\partial}{\partial x} + \mathrm{i}cm\omega x \\ c\frac{\hbar}{\mathrm{i}}\frac{\partial}{\partial x} - \mathrm{i}cm\omega x & -mc^2 \end{pmatrix} \tag{2.29}$$

we write

$$\Psi(x) = \begin{pmatrix} \psi_1(x) \\ \psi_2(x) \end{pmatrix} \tag{2.30}$$

we can solve the second line for ψ_2,

$$\psi_2 = \frac{c\frac{\hbar}{i}\psi_1' - icm\omega x\psi_1}{E + mc^2} \; . \tag{2.31}$$

Inserting into the first line we obtain the following equation for ψ_1,

$$\left(\frac{E^2 - m^2c^4}{2mc^2} + \frac{\hbar\omega}{2}\right)\psi_1 = -\frac{\hbar^2}{2m}\psi_1'' + \frac{m}{2}\omega^2 x^2 \psi_1 \; , \tag{2.32}$$

which formally is the equation for a harmonic oscillator, cf. (2.1), and thus the energy eigenvalues are given by

$$E_n^\pm = \pm\sqrt{m^2c^4 + 2\hbar\omega mc^2 n} \; , \quad n \in \mathbb{N}_0 \; . \tag{2.33}$$

Here we see that in the non-relativistic limit, i.e. for energies close to the rest energy mc^2, the system reduces to a harmonic oscillator with energies $E_n^\pm \mp mc^2 = \pm\hbar\omega n + \mathcal{O}\left(c^{-2}\right)$. From (2.33) we now obtain a trace formula by Poisson summation as in the previous cases,

$$\begin{aligned} d(E) &= \sum_{n=0}^{\infty} \left[\delta(E - E_n^+) + \delta(E - E_n^-)\right] \\ &= \sum_{k\in\mathbb{Z}} \int_0^\infty \left[\delta(E - E_n^+) + \delta(E - E_n^-)\right] \mathrm{e}^{2\pi i k n} \, \mathrm{d}n \\ &= \begin{cases} \displaystyle\sum_{k\in\mathbb{Z}} \frac{|E|}{\hbar\omega mc^2} \exp\left(-\pi i k \frac{E^2 - m^2c^4}{\hbar\omega mc^2}\right) & , \; |E| \geq mc^2 \\ 0 & , \; \text{otherwise} \end{cases} \end{aligned} \tag{2.34}$$

Before we can relate this result to classical objects we have to say some words about semiclassical approximations for Dirac operators in order to define the corresponding classical dynamics. To this end consider the WKB ansatz

$$\Psi(x) = (a_0(x) + \hbar a_1(x) + \ldots)\, \mathrm{e}^{(i/\hbar)S(x)} \tag{2.35}$$

with a scalar phase S and and spinor valued amplitudes a_j. Inserting into the stationary Dirac equation with Hamiltonian (2.29) and sorting by powers of $\hbar$, in leading semiclassical order we obtain

$$\begin{pmatrix} -E + mc^2 & c\frac{\partial S}{\partial x}(x) + icm\omega x \\ c\frac{\partial S}{\partial x}(x) - icm\omega x & -E - mc^2 \end{pmatrix} a_0(x) = 0 \; . \tag{2.36}$$

In order to find a non-trivial solution of this condition the matrix in (2.36) must have an eigenvalue zero, i.e.

$$-E \pm \sqrt{m^2c^4 + c^2\left(\frac{\partial S}{\partial x}\right)^2 + c^2m^2\omega^2x^2} = 0\,. \tag{2.37}$$

If we interpret these equations as Hamilton–Jacobi equations, as in the ordinary WKB-method, we find the classical Hamiltonians

$$H^{\pm}(p,x) = \pm\sqrt{m^2c^4 + c^2p^2 + c^2m^2\omega^2x^2} \tag{2.38}$$

corresponding to positive and negative energies, respectively. Trajectories in phase space follow the energy contour $H^{\pm}(p,x) = E$ which is an ellipse (as for the harmonic oscillator) with normal form

$$1 = \frac{p^2}{\frac{E^2 - m^2c^4}{c^2}} + \frac{x^2}{\frac{E^2 - m^2c^4}{c^2m^2\omega^2}}\,. \tag{2.39}$$

The action of the (primitive) periodic orbit is the area of the ellipse,

$$S = \oint p\,\mathrm{d}x = \pi\frac{E^2 - m^2c^4}{c^2m\omega} \tag{2.40}$$

multiples of which are readily identified with the exponents in (2.34).

The Maslov index of the ellipse is two as in the case of the harmonic oscillator. However, now we face a puzzle, as it seems that the Maslov phase is missing in (2.34). The solution is hidden in the fact that in order to find solutions of the higher order equations arising from the ansatz (2.35) these equations first have to be projected onto the subspaces corresponding to the classical Hamiltonians (2.38). The next-to-leading order equation (the terms proportional to $\hbar^1$) is needed to determine the amplitudes in the trace formula. When doing so one finds an additional weight factor similar to the factor $\cos(k\alpha/2)$ in the previous example (2.21), which here is a simple phase factor that exactly cancels the Maslov phase. In the general case of the Dirac equation with arbitrary electro-magnetic fields, see Sect. 3.7 and [9, 10], the analogous procedure leads to a classical equation of spin precession of the form (2.22) which has to be integrated along classical relativistic trajectories. In this sense the phase factor which here cancels the Maslov phase is a remnant of the classical spin precession, although, there is no spin degree of freedom included in the model (2.29), since the two components correspond to particles and antiparticles rather than to spin up and spin down.

Note that this phase factor also yields the difference of the spectrum (2.33) and the collection of the spectra of the pseudo-differential operators with Weyl symbols (2.38) for which by EBK-quantisation we find the eigenvalues

$$E_n^{\pm} = \pm\sqrt{m^2c^4 + 2\hbar\omega mc^2\left(n + \tfrac{1}{2}\right)}\,, \quad n \in \mathbb{N}_0\,. \tag{2.41}$$

Notice that one also obtains the latter energy spectrum when investigating the stationary Klein–Gordon equation

$$\left(-c^2\hbar^2 \frac{\partial^2}{\partial x^2} + m^2c^4 + c^2m^2\omega^2x^2 \right) \psi(x) = E^2 \psi(x) , \tag{2.42}$$

which formally is again the equation of a harmonic oscillator, cf. (2.1). Therefore, in this sense, the additional phase factor, which cancels the Maslov phase in (2.34), distinguishes between relativistic Fermions and Bosons in a harmonic oscillator potential.

We remark that the same effect as for the Dirac oscillator also rendered Sommerfeld's relativistic theory of the fine structure of the hydrogen atom [13, 14] exact. Having performed his calculation within the old quantum theory predating the invention of quantum mechanics and the discovery of spin, he could neither know of the Maslov correction nor of the spin correction. By a freak of nature these contributions also cancel in the case of the hydrogen atom as we will show in Sect. 5.6, which allowed Sommerfeld to find the exact eigenvalues of the hydrogen atom with an electron with spin 1/2 more than 10 years before the Dirac equation was invented.

Summarising this short overview of trace formulae with spin, we notice that spin yields additional weight factors which lead to corrections of the same order as the Maslov corrections, and thus, can neither be neglected in trace formulae nor in semiclassical quantisation conditions.

2.2 Spectral Statistics

Having explored the structure of semiclassical trace formulae we now introduce some spectral functions and investigate how trace formulae and thus classical properties can be used in order to extract information on the quantum mechanical spectral statistics. We will again use the model systems (various oscillators) introduced in the previous section for which the statistical information is easily obtained directly from the explicitly known spectra. However, it is instructive to express this information in terms of classical objects which, e.g., allows a particularly clear introduction to the concept of the diagonal approximation which will play an important role in Chap. 7.

The concept of universality in spectral statistics [15, 16] asserts that the fluctuations in the distribution of eigenvalues is related to the integrability or chaoticity of the corresponding classical system. In order to be able to compare these fluctuations for different systems the spectra have to be unfolded, which means that instead of the eigenenergies E_n we consider the unfolded energies $x_n := f(E_n)$, where the function f is chosen such that the mean spacing of the new spectrum $\{x_n\}$ is rescaled to unity. This can, e.g., be achieved by defining

$$x_n := \bar{N}(E_n) , \tag{2.43}$$

where $\bar{N}(E)$ is the non-oscillating part of the trace formula for the integrated spectral density $N(E)$, cf. (2.27). For the harmonic oscillator we have $\bar{N}(E) = 1/(\hbar\omega)$ which together with the eigenvalues (2.2) yields

$$x_n = n + \frac{1}{2} . \tag{2.44}$$

For the harmonic oscillator with spin 1/2 there are on average twice as many eigenvalues E_n, yielding $\bar{N}(E) = 2/(\hbar\omega)$ and thus with (2.18) we obtain the unfolded spectrum

$$x_n^{\pm} = 2n + 1 \pm \frac{g}{\hbar\omega}\mu_B|\boldsymbol{B}| = 2n + 1 \pm \varepsilon . \tag{2.45}$$

Our last example is the Dirac oscillator for which the spectrum is symmetric with respect to $E = 0$, cf. (2.33). Therefore, it is sufficient to consider the positive part $\{E_n^+\}$. From (2.34) we have $\bar{d} = E/(\hbar\omega mc^2)$, $E \geq mc^2$, and thus $\bar{N}(E) = \int_{mc^2}^{E} \bar{d}(E')\,\mathrm{d}E' = (E^2 - m^2c^4)/(2\hbar\omega mc^2)$, yielding

$$x_n = n , \tag{2.46}$$

which, apart from a trivial translation by 1/2 that does not influence the spectral correlations, is identical to the unfolded spectrum (2.44) of the harmonic oscillator.

For convenience we also define the density of states $\mathscr{d}(x)$, $x = \bar{N}(E)$, of the unfolded spectrum,

$$\begin{aligned}\mathscr{d}(x) :&= \sum_n \delta(x - x_n) = \sum_n \delta(\bar{N}(E) - \bar{N}(E_n)) = \sum_n \frac{1}{\bar{d}(E)}\delta(E - E_n) \\ &= d(E)/\bar{d}(E) ,\end{aligned} \tag{2.47}$$

and its spectral staircase function

$$\mathscr{N}(x) := \int_0^x \mathscr{d}(x')\,\mathrm{d}x' = N(E) . \tag{2.48}$$

First, we want to investigate two-point correlations which can, e.g., be done in terms of the two-point correlation function

$$R_2(s) := \langle \mathscr{d}(x+s)\,\mathscr{d}(x)\rangle - 1 , \tag{2.49}$$

where $\langle\,\cdot\,\rangle$ denotes a suitable average over x. For semiclassical investigations it is often more convenient to consider instead the spectral form factor

$$K(\tau) := \int_{\mathbb{R}} R_2(s)\,\mathrm{e}^{-2\pi i s\tau}\,\mathrm{d}s , \tag{2.50}$$

the Fourier transform of $R_2(s)$. Since both the two-point correlation function and the spectral form factor obviously are distributions which have to be evaluated on smooth test functions, it is also common to investigate two-point correlations in terms of statistic functions which can be expressed as a convolution of either $R_2(s)$ or $K(\tau)$ with some integral kernel. A particular example is the number variance

$$\Sigma^2(L) := \left\langle (\mathcal{N}(x+L) - \mathcal{N}(x) - L)^2 \right\rangle \tag{2.51}$$

which measures the deviation of the number of levels in an interval from the mean number given by the length L of that interval. The number variance can be expressed as

$$\Sigma^2(L) = \frac{2}{\pi^2} \int_0^\infty K(\tau) \frac{\sin^2(\pi\tau L)}{\tau^2} \, \mathrm{d}\tau \;, \tag{2.52}$$

see e.g. [17].

Let us now consider the two-point correlation function for the harmonic oscillator (or for the Dirac oscillator, which is the same as we noticed earlier). The spectrum is equidistant and therefore the two-point correlation function is given by a delta comb,

$$R_2(s) = \sum_{n \in \mathbb{Z}} \delta(s-n) - 1 \;. \tag{2.53}$$

Thus, the form factor is also given by a delta comb,

$$K(\tau) = \sum_{n \in \mathbb{Z}} \mathrm{e}^{-2\pi i n \tau} - \delta(\tau) = \sum_{n \in \mathbb{Z} \setminus \{0\}} \delta(\tau - n) \;. \tag{2.54}$$

So far this is trivial, however, it is now interesting to see how the trace formula reproduces these results. From (2.13) and (2.47) we obtain

$$d(x) = 1 + \sum_{k \in \mathbb{Z} \setminus \{0\}} \mathrm{e}^{(\mathrm{i}/\hbar) k S - \mathrm{i}\pi k} \;, \tag{2.55}$$

where the action S (2.10) expressed in terms of the unfolded energy variable $x = \bar{N}(E) = E/(\hbar\omega)$ is given by

$$S = TE = \hbar\omega T x = 2\pi\hbar\, x \;. \tag{2.56}$$

Substituting into the definition (2.49) we obtain,

$$\begin{aligned} R_2(s) = & \left\langle \sum_{k \in \mathbb{Z} \setminus \{0\}} \mathrm{e}^{2\pi \mathrm{i} k x - \mathrm{i}\pi k} \right\rangle + \left\langle \sum_{k \in \mathbb{Z} \setminus \{0\}} \mathrm{e}^{2\pi \mathrm{i} k (x+s) - \mathrm{i}\pi k} \right\rangle \\ & + \left\langle \sum_{k \in \mathbb{Z} \setminus \{0\}} \sum_{k' \in \mathbb{Z} \setminus \{0\}} \mathrm{e}^{2\pi \mathrm{i} k x - \mathrm{i}\pi k} \, \mathrm{e}^{2\pi \mathrm{i} k' (x+s) - \mathrm{i}\pi k'} \right\rangle . \end{aligned} \tag{2.57}$$

Usually one would choose the energy average $\langle \cdot \rangle$ such that many eigenvalues are included, but since the spectrum under consideration is periodic we can choose $\langle\,\cdot\,\rangle = \int_0^1 \cdot\, \mathrm{d}x$, for simplicity. We immediately see that the first two terms in (2.57) vanish and thus $R_2(s)$ is given by the correlator of the periodic orbit sums. This result is also true in general, since the first two terms are averages of periodic orbit sums over some energy range, where the sums themselves fluctuate about zero.

Investigating the last term of (2.57) we change variables from k' to $-k'$ obtaining

$$R_2(s) = \sum_{k\in\mathbb{Z}\setminus\{0\}} \sum_{k'\in\mathbb{Z}\setminus\{0\}} \left\langle \mathrm{e}^{2\pi\mathrm{i}(k-k')x} \right\rangle \mathrm{e}^{-2\pi\mathrm{i}k's}\, \mathrm{e}^{-\mathrm{i}\pi(k-k')}\,, \tag{2.58}$$

where, due to the energy average, only the terms with $k = k'$ contribute. This selection of the diagonal terms in the double periodic orbit sum, which here is exact, is known as the diagonal approximation. In Sect. 7.3.1 we will see that the term corresponding to $\exp(2\pi\mathrm{i}(k-k')x)$ in general oscillates with a frequency proportional to the difference $T - T'$ of the periods of two periodic orbits of the classical system. Thus, one obtains a large contribution for the correlator of identical orbits (or of pairs which are related by a symmetry, like time reversal), but in general it is not obvious whether the off-diagonal terms may be neglected, since usually no lower bounds are known for $T - T'$ with T and T' corresponding to different orbits. A more detailed discussion of this topic will be given in Sect. 7.3.1.

In the present situation, however, the diagonal approximation is exact and the semiclassics reproduces the exact quantum mechanical two-point correlation function (2.53). Clearly, this result also holds for the Fourier transform, the spectral form factor $K(\tau)$.

We now want to discuss the number variance $\Sigma^2(L)$. To this end let us turn to the slightly more complicated spectrum (2.45) of the harmonic oscillator with spin. Using again the dimensionless parameter $\varepsilon := \frac{g}{\hbar\omega}\mu_B|\boldsymbol{B}|$ the unfolded eigenvalues alternately come at distances 2ε and $1-2\varepsilon$. Since we are still dealing with a periodic spectrum the average in the definition (2.51) can be chosen as $\langle\,\cdot\,\rangle = \frac{1}{2}\int_0^2 \cdot\, \mathrm{d}x$. Furthermore, we can choose $\varepsilon < \frac{1}{2}$ without restriction. The periodicity of $d(x)$ now forces $\Sigma^2(L)$ to also be periodic with period 2. Thus it is sufficient to consider $0 < L < 2$.

The difference $\Delta\mathcal{N} := \mathcal{N}(x+L) - \mathcal{N}(x)$ counts the number of eigenvalues in the interval $(x, x+L)$. For $L < 2\varepsilon$ we can either find $\Delta\mathcal{N} = 0$ or $\Delta\mathcal{N} = 1$ but never $\Delta\mathcal{N} > 1$. Since the mean density is one, the probability of finding $\Delta\mathcal{N} = 1$ is given by $p(\Delta\mathcal{N} = 1) = L$, and by normalisation we conclude that $p(\Delta\mathcal{N} = 0) = 1 - L$. Therefore, for $L < 2\varepsilon$ we have

$$\Sigma^2(L) = (0-L)^2\,(1-L) + (1-L)^2\,L = L - L^2\,. \tag{2.59}$$

For the range $2\varepsilon < L < 2 - 2\varepsilon$ we find the average number of eigenvalues in the interval $(x, x+L)$ by direct computation,

$$\Delta\mathcal{N} = \begin{cases} 0 \, , & 0 < x < 1 - \varepsilon - L \quad \text{or} \quad 1 + \varepsilon < x < 2 \\ 1 \, , & 1 - \varepsilon - L < x < 1 + \varepsilon - L \quad \text{or} \quad 1 - \varepsilon < x < 1 + \varepsilon \\ 2 \, , & 1 + \varepsilon - L < x < 1 - \varepsilon \end{cases} , \tag{2.60}$$

yielding $\Sigma^2(L) = -2\varepsilon + 2L - L^2$. Analogously we find the result for $L > 2 - 2\varepsilon$ and altogether the number variance is given by

$$\Sigma^2(L) = \begin{cases} L - L^2 \quad , & L < 2\varepsilon \\ -2\varepsilon + 2L - L^2 \, , & 2\varepsilon < L < 2 - 2\varepsilon \\ -2 + 3L - L^2 \, , & L > 2 - 2\varepsilon \end{cases} , \tag{2.61}$$

for $0 < L < 2$ and continued periodically for $L > 2$.

For comparison we now calculate the number variance using the trace formula (2.27). With (2.48) and $x = \bar{N}(E) = 2E/(\hbar\omega)$ we find

$$\mathcal{N}(x + L) - \mathcal{N}(x) - L = \sum_{k \in \mathbb{Z} \setminus \{0\}} 2\cos(k\pi\epsilon)\, \mathrm{e}^{\mathrm{i}k\pi(x-1)} \left(\mathrm{e}^{\mathrm{i}k\pi L} - 1\right) . \tag{2.62}$$

When substituting this expression into the definition (2.51) the energy average again cancels all terms except for the diagonal terms, yielding

$$\Sigma^2(L) = \sum_{k=1}^{\infty} \left(\frac{2}{\pi k}\right)^2 \cos^2(k\pi\epsilon)\,(1 - \cos(k\pi L)) \, . \tag{2.63}$$

Here we see that the number variance oscillates about the average value

$$\Sigma^2_\infty := \sum_{k=1}^{\infty} \left(\frac{2}{\pi k}\right)^2 \cos^2(k\pi\epsilon) \, . \tag{2.64}$$

We can now investigate what happens if we approximate the number variance by cutting off the k-sum in (2.63) at some value K, i.e. in the trace formula we only take into account (repetitions of) orbits up to a certain period. Since as a function of energy E the individual terms in the trace formula oscillate on scales proportional to the inverse periods, this cut-off is equivalent to observing the quantum mechanical energy spectrum with a finite resolution. Doing so not only smoothes the number variance $\Sigma^2(L)$ at discontinuities but by (2.64) also lowers the mean value $\Sigma^2_\infty(L)$. This behaviour is illustrated in Fig. 2.3, where we show the exact number variance (2.61) compared to approximations based on (2.63) using different cut-offs K. In the general case a similar behaviour of $\Sigma^2(L)$ is observed for large values of L, known as the saturation regime. There the number variance deviates significantly from the curves expected according to the Bohigas–Giannoni–Schmit or Berry–Tabor conjectures. This effect can also be related to the finite resolution with which the quantum spectrum is known. The relevant scale on which saturation sets in is determined by the ratio of the periods of short periodic orbits and the Heisenberg time $T_H := 2\pi\hbar\bar{d}$ which is the time scale related to the mean spectral density. Similarly, the saturation value itself is determined by the cut-off and the Heisenberg time as will be explained in Sect. 7.5.1.

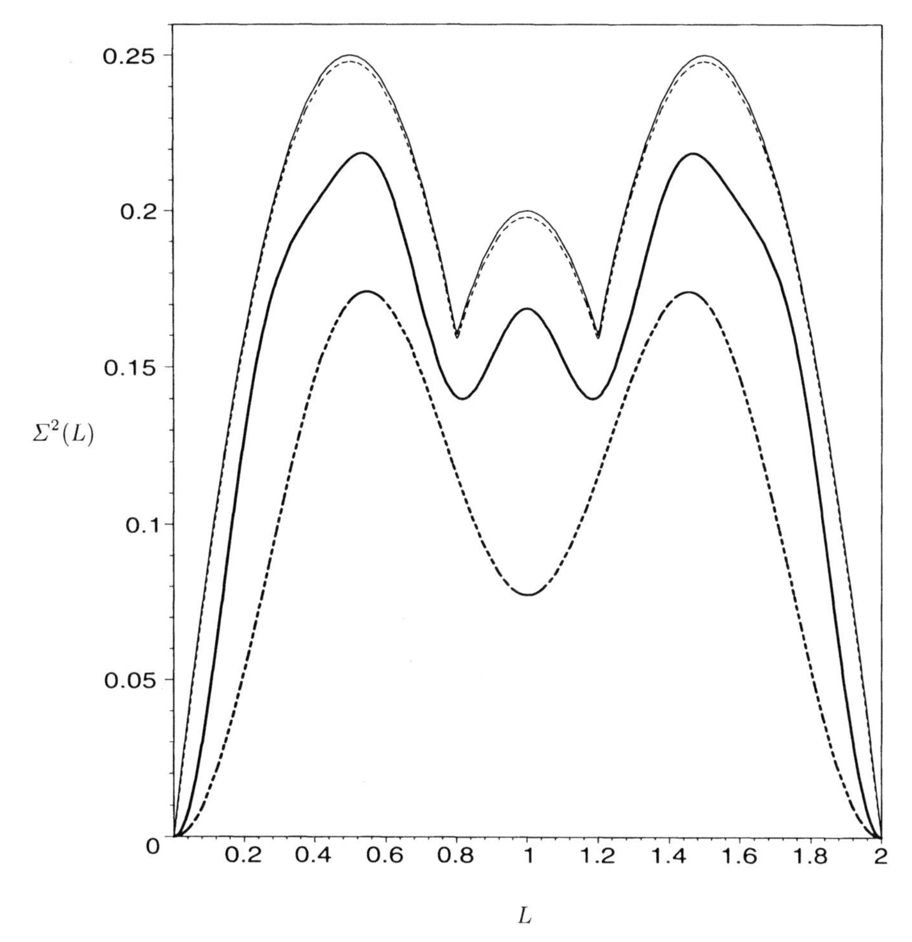

Fig. 2.3. The exact number variance (2.61) with $\varepsilon = 0.4$ (*solid line*) compared to the semiclassical expression (2.63) with cut-offs $K = 100$ (*dashed line*), $K = 5$ (*thick solid line*) and $K = 2$ (*thick dashed line*), respectively.

References

1. A. Messiah: *Quantum Mechanics. Vol. I*, Translated from the French by G. M. Temmer (North-Holland Publishing Co., Amsterdam, 1961)
2. M.C. Gutzwiller: J. Math. Phys. **12**, 343–358 (1971)
3. M.C. Gutzwiller: *Chaos in Classical and Quantum Mechanics* (Springer-Verlag, New York, 1990)
4. M.C. Gutzwiller: J. Math. Phys. **11**, 1791–1806 (1970)
5. M.V. Berry, M. Tabor: Proc. R. Soc. London Ser. A **349**, 101–123 (1976)
6. M.V. Berry, M. Tabor: J. Phys. A **10**, 371–379 (1977)
7. L.H. Thomas: Philos. Mag. **3**, 1–22 (1927)
8. V. Bargman, L. Michel, V.L. Telegdi: Phys. Rev. Lett. **2**, 435–436 (1959)

9. J. Bolte, S. Keppeler: Phys. Rev. Lett. **81**, 1987–1991 (1998)
10. J. Bolte, S. Keppeler: Ann. Phys. (NY) **274**, 125–162 (1999)
11. M. Moshinsky, A. Szczepaniak: J. Phys. A **22**, L817–L819 (1989)
12. P. Strange: *Relativistic Quantum Mechanics with Applications in Condensed Matter and Atomic Physics* (Cambridge University Press, Cambridge, 1998)
13. A. Sommerfeld: Ann. Phys. (Leipzig) **51**, 1–94, 125–167 (1916)
14. A. Sommerfeld: *Atombau und Spektrallinien*, Vol. I, 8th edn. (Friedr. Viehweg & Sohn GmbH, Braunschweig, 1969)
15. M.V. Berry, M. Tabor: Proc. R. Soc. London Ser. A **356**, 375–394 (1977)
16. O. Bohigas, M.J. Giannoni, C. Schmit: Phys. Rev. Lett. **52**, 1–4 (1984)
17. M.L. Mehta: *Random Matrices*, 2nd edn. (Academic Press, San Diego, 1991)

3 Trace Formulae with Spin

This chapter is dedicated to the derivation of semiclassical trace formulae for particles with spin. We will focus on Pauli Hamiltonians with arbitrary spin $s \in \mathbb{N}/2$ and only briefly describe the corresponding calculations for the Dirac equation. We will also comment on a different limit in which the spectral asymptotics of the Pauli Hamiltonian can be considered. Semiclassical trace formulae for the Dirac equation and its non-relativistic limit, the Pauli equation with spin 1/2, are discussed in detail in [1–3]; trace formulae with observables are discussed in [4], and a brief overview can also be found in [5].

The theory of semiclassical trace formulae, which asymptotically (in the limit $\hbar \to 0$) express the quantum mechanical density of states in terms of a classical periodic orbit sum, was developed by Gutzwiller around 1970 in a series of articles [6–9]. Apparently, a parallel development can be found in the work of Balian and Bloch [10, 11]. At about the same time also in the mathematical literature there appeared various articles exploring the relation between the quantum spectrum and classical periodic orbits [12–15]. These authors explore the relation in the high energy limit rather than in the semiclassical limit $\hbar \to 0$. The methods used in this context are known under the notion of microlocal analysis, see e.g. [16–18]. Derivations of the Gutzwiller trace formula (i.e. asymptotics in the limit $\hbar \to 0$) in the setting of microlocal analysis can be found in [19, 20]. Today there exist various monographs focusing on different aspects of semiclassical trace formulae and related topics; a selection is given by [21–27].

3.1 The Pauli Hamiltonian

The Pauli equation describing non-relativistic particles with spin s in d spatial dimensions reads

$$\mathrm{i}\hbar \frac{\partial}{\partial t} \Psi(\boldsymbol{x}, t) = \hat{H}_{\mathrm{P}} \Psi(\boldsymbol{x}, t) , \tag{3.1}$$

where Ψ is a (2s+1) spinor, i.e. $\Psi \in L^2(\mathbb{R}^d) \otimes \mathbb{C}^{2s+1}$. The Pauli Hamiltonian $\hat{H}_{\mathrm{P}}$ is of the form

$$\hat{H}_{\mathrm{P}} = \hat{H} \mathbb{1}_{2s+1} + \hat{\boldsymbol{s}} \hat{\boldsymbol{\mathcal{B}}} , \tag{3.2}$$

where $\hat{H}$ is a Schrödinger operator with Weyl symbol (cf. Appendix E) $H(\boldsymbol{p}, \boldsymbol{x})$ and $\mathbb{1}_{2s+1}$ denotes the $(2s+1) \times (2s+1)$ unit matrix. For instance, for a particle with mass m and charge e in external electro-magnetic fields defined by the potentials $(\phi, \boldsymbol{A})$ the symbol of $\hat{H}$ reads

$$H(\boldsymbol{p}, \boldsymbol{x}) = \frac{1}{2m}\left(\boldsymbol{p} - \frac{e}{c}\boldsymbol{A}(\boldsymbol{x})\right)^2 + e\phi(\boldsymbol{x}) \,. \tag{3.3}$$

The second term in (3.2) describes the coupling of the spin operator $\hat{\boldsymbol{s}}$ to some vector $\hat{\boldsymbol{\mathcal{B}}}$ of Weyl operators, representing, for instance, the influence of an external magnetic field or spin–orbit coupling. In these two cases (for an electron) the Weyl symbol $\boldsymbol{\mathcal{B}}$ is given by

$$\begin{aligned} \boldsymbol{\mathcal{B}}_{\text{mag}}(\boldsymbol{p}, \boldsymbol{x}) &= -\frac{e}{mc}\boldsymbol{B}(\boldsymbol{x}) \,, \quad \text{with magnetic field } \boldsymbol{B}(\boldsymbol{x}) \text{ and} \\ \boldsymbol{\mathcal{B}}_{\text{so}}(\boldsymbol{p}, \boldsymbol{x}) &= \frac{e}{2m^2c^2}\,\boldsymbol{p} \times \boldsymbol{E}(\boldsymbol{x}) \,, \quad \text{with electric field } \boldsymbol{E}(\boldsymbol{x}), \end{aligned} \tag{3.4}$$

respectively. The spin operator is defined via the derived $(2s+1)$-dimensional representation of $\mathfrak{su}(2)$ by

$$\hat{\boldsymbol{s}} := \frac{\hbar}{2}\mathrm{d}\pi_s(\boldsymbol{\sigma}) \,, \tag{3.5}$$

where π_s denotes the $(2s+1)$-dimensional[1] (unitary, irreducible) representation of SU(2) and

$$\mathrm{d}\pi_s(X) = \frac{1}{\mathrm{i}}\frac{\mathrm{d}}{\mathrm{d}\lambda}\pi_s\left(\mathrm{e}^{\mathrm{i}\lambda X}\right)\Big|_{\lambda=0} \tag{3.6}$$

for any $X \in \mathfrak{su}(2)$, see also Appendix C for details. E.g., for a spin 1/2 we have

$$\mathrm{d}\pi_{1/2}(\boldsymbol{\sigma}) = \boldsymbol{\sigma} \tag{3.7}$$

with the Pauli matrices

$$\sigma_x = \begin{pmatrix} 0 & 1 \\ 1 & 0 \end{pmatrix} , \quad \sigma_y = \begin{pmatrix} 0 & -\mathrm{i} \\ \mathrm{i} & 0 \end{pmatrix} \quad \text{and} \quad \sigma_z = \begin{pmatrix} 1 & 0 \\ 0 & -1 \end{pmatrix} , \tag{3.8}$$

and thus the corresponding spin operator reads

$$\hat{\boldsymbol{s}} = \frac{\hbar}{2}\boldsymbol{\sigma} \,. \tag{3.9}$$

Analogously, for a spin 1 we find

$$\mathrm{d}\pi_1(\boldsymbol{\sigma}) = \left(\begin{pmatrix} 0 & \sqrt{2} & 0 \\ \sqrt{2} & 0 & \sqrt{2} \\ 0 & \sqrt{2} & 0 \end{pmatrix} , \begin{pmatrix} 0 & -\mathrm{i}\sqrt{2} & 0 \\ \mathrm{i}\sqrt{2} & 0 & -\mathrm{i}\sqrt{2} \\ 0 & \mathrm{i}\sqrt{2} & 0 \end{pmatrix} , \begin{pmatrix} 2 & 0 & 0 \\ 0 & 0 & 0 \\ 0 & 0 & -2 \end{pmatrix} \right)^T \,. \tag{3.10}$$

Since the spin operator (3.5) is proportional to $\hbar$, the Pauli operator (3.2) can be viewed as a semiclassical expansion of some Hamiltonian

[1] Note that we use the spin quantum number s rather than the dimension $(2s+1)$ as an index to characterise the representations of SU(2), i.e. we have $\pi_s(id) = \mathbb{1}_{2s+1}$.

$$H(\hat{\boldsymbol{p}}, \hat{\boldsymbol{x}}, \hat{\boldsymbol{s}}) = H_0(\hat{\boldsymbol{p}}, \hat{\boldsymbol{x}}) + \boldsymbol{\mathcal{B}}(\hat{\boldsymbol{p}}, \hat{\boldsymbol{x}})\hat{\boldsymbol{s}} + \mathcal{O}\left(\hbar^2\right) , \tag{3.11}$$

where H_0 describes the translational dynamics and $\boldsymbol{\mathcal{B}}$ the coupling to the spin degree of freedom to lowest order as $\hbar \to 0$. Since for the semiclassical questions to be discussed, higher order terms are irrelevant we may without loss of generality restrict ourselves to the study of Hamiltonians of the form (3.2) which are linear in $\hat{\boldsymbol{s}}$. In addition, it is exactly this type of operator which one obtains from a non-relativistic approximation of the Dirac equation, see e.g. [28, 29]. On the other hand, if in a semiclassical study one chooses to ignore the $\hbar$-dependence of the spin operator $\hat{\boldsymbol{s}}$, see e.g. [30, 31], an operator describing a spin-s particle can be an arbitrary (hermitian) $(2s+1)\times(2s+1)$-matrix valued operator. We will briefly return to this topic in Sect. (3.8).

3.2 Deriving Trace Formulae: General Strategy

Here we want to give a brief introduction on how to derive (regularised) semiclassical trace formulae in general. The starting point is always a semiclassical representation of the time evolution kernel $K(\boldsymbol{x}, \boldsymbol{y}, t)$, which propagates an initial state $\Psi(\boldsymbol{x}, 0)$ to a state at time t by

$$\Psi(\boldsymbol{x}, t) = \int_{\mathbb{R}^d} K(\boldsymbol{x}, \boldsymbol{y}, t)\, \Psi(\boldsymbol{y}, 0)\, \mathrm{d}^d y . \tag{3.12}$$

Notice that K takes values in the unitary $(2s + 1) \times (2s + 1)$ matrices. For (3.12) to be consistent we have to impose the initial condition

$$K(\boldsymbol{x}, \boldsymbol{y}, 0) = \delta^{(d)}(\boldsymbol{x} - \boldsymbol{y}) . \tag{3.13}$$

Clearly, for Ψ to be a solution of the Pauli equation, K also has to solve

$$\left(\hat{H}_{\mathrm{P}} - \mathrm{i}\hbar\frac{\partial}{\partial t}\right) K(\boldsymbol{x}, \boldsymbol{y}, t) = 0 , \tag{3.14}$$

where $\hat{H}_{\mathrm{P}}$ acts on the first argument of K. The semiclassical time evolution kernel for the Schrödinger equation ($s = 0$) is known as the Van Vleck formula or the Van Vleck–Gutzwiller propagator, cf. [6, 7, 32]; see also [33] for a historical account on the development of the formula.

On the other hand, the time evolution kernel has a spectral representation, which, in case of a purely discrete spectrum, with eigenvalues E_n and eigenstates φ_n, reads

$$K(\boldsymbol{x}, \boldsymbol{y}, t) = \sum_n \varphi_n(\boldsymbol{x})\, \varphi_n^\dagger(\boldsymbol{y})\, \mathrm{e}^{-(\mathrm{i}/\hbar)E_n t} , \tag{3.15}$$

where the dagger "†" denotes hermitian conjugation. We remark that the restriction to a purely discrete spectrum was chosen in order to simplify the presentation; using the methods of [22], in [3] it is shown that all results are still valid in the presence of a continuous spectrum which is well separated from the discrete spectrum.

In order to obtain the spectral density one now wants to change from the time domain to the energy domain by a Fourier transform, yielding the Green's function of the time independent wave equation, and then take the trace. However, instead of dealing with distributions we prefer to use convergent sums. To this end we first multiply (3.15) with a smooth test function $\hat{\varrho}(t) \in C_0^\infty(\mathbb{R})$ being the Fourier transform

$$\hat{\varrho}(t) = \int_{\mathbb{R}} \varrho(\omega)\, \mathrm{e}^{\mathrm{i}\omega t}\, \mathrm{d}\omega \tag{3.16}$$

of an equally smooth function $\varrho(\omega)$; more precisely $\varrho, \hat{\varrho} \in \mathcal{S}(\mathbb{R})$. Only then we calculate Fourier transform and trace. Thus, we have to evaluate the expression

$$\frac{1}{2\pi} \int_{\mathbb{R}} \hat{\varrho}(t)\, [\mathrm{Tr}\, K](t)\, \mathrm{e}^{(\mathrm{i}/\hbar)Et}\, \mathrm{d}t = \mathrm{tr} \int_{\mathbb{R}^d} \frac{1}{2\pi} \int_{\mathbb{R}} \hat{\varrho}(t)\, K(\boldsymbol{x}, \boldsymbol{x}, t)\, \mathrm{e}^{(\mathrm{i}/\hbar)Et}\, \mathrm{d}t\, \mathrm{d}^d x\ , \tag{3.17}$$

where by Tr we denote the trace operation on the full Hilbert space $L^2(\mathbb{R}^d) \otimes \mathbb{C}^{2s+1}$ as opposed to the matrix trace which we will always denote by tr. Substituting (3.15) and (3.16) we find

$$\begin{aligned}
&\frac{1}{2\pi} \int_{\mathbb{R}} \hat{\varrho}(t)\, [\mathrm{Tr}\, K](t)\, \mathrm{e}^{(\mathrm{i}/\hbar)Et}\, \mathrm{d}t \\
&\quad = \int_{\mathbb{R}^d} \int_{\mathbb{R}} \sum_n \underbrace{\frac{1}{2\pi} \int_{\mathbb{R}} \mathrm{e}^{\mathrm{i}(\omega - (E_n - E)/\hbar)t}\, \mathrm{d}t}_{\delta\left(\omega - \frac{E_n - E}{\hbar}\right)}\, \varrho(\omega)\, \underbrace{\mathrm{tr}\left[\varphi_n(\boldsymbol{x})\, \varphi_n^\dagger(\boldsymbol{x})\right]}_{\varphi_n^\dagger(\boldsymbol{x})\, \varphi_n(\boldsymbol{x})}\, \mathrm{d}\omega\, \mathrm{d}^d x \\
&\quad = \sum_n \varrho\left(\frac{E_n - E}{\hbar}\right) \int_{\mathbb{R}^d} \varphi_n^\dagger(\boldsymbol{x})\, \varphi_n(\boldsymbol{x})\, \mathrm{d}^d x \\
&\quad = \sum_n \varrho\left(\frac{E_n - E}{\hbar}\right) .
\end{aligned} \tag{3.18}$$

A trace formula for the spectral function (3.18) is then obtained by inserting a semiclassical expression for the time evolution kernel, which we will explicitly determine in the following sections. In this section we only want to list some basic properties. A semiclassical ansatz for the time evolution operator can always be made in terms of an oscillatory integral,

$$K(\boldsymbol{x}, \boldsymbol{y}, t) = \int_{\mathbb{R}^d} [a_0(\boldsymbol{x}, \boldsymbol{\xi}, t) + \hbar a_1(\boldsymbol{x}, \boldsymbol{\xi}, t) + \ldots]\, \mathrm{e}^{(\mathrm{i}/\hbar)(S(\boldsymbol{x}, \boldsymbol{\xi}, t) - \boldsymbol{y}\boldsymbol{\xi})}\, \mathrm{d}^d \xi\ , \tag{3.19}$$

with a scalar function S and matrix-valued amplitudes a_k. Later we will justify this ansatz and determine the leading order behaviour, which is obtained by calculating S and a_0 and performing the $\boldsymbol{\xi}$-integral with the method of stationary phase. For the purpose of this first discussion it is sufficient to know that it will turn out that the phase S has to solve a Hamilton–Jacobi equation with classical Hamiltonian $H(\boldsymbol{p},\boldsymbol{x})$, cf. (3.2). Thus $S(\boldsymbol{x},\boldsymbol{\xi},t)$ generates the classical motion from $(\boldsymbol{\xi},\nabla_{\boldsymbol{\xi}}S)$ to $(\nabla_{\boldsymbol{x}}S,\boldsymbol{x})$ in time t. The stationary phase condition in (3.19) reads

$$\nabla_{\boldsymbol{\xi}} S(\boldsymbol{x},\boldsymbol{\xi},t) = \boldsymbol{y} \,, \tag{3.20}$$

i.e., $K(\boldsymbol{x},\boldsymbol{y},t)$ is given by a sum over classical trajectories connecting $\boldsymbol{y}$ and $\boldsymbol{x}$ in time t. Moreover, a semiclassical expansion of (3.18) requires the stationarity of the total phase

$$\phi(\boldsymbol{x},\boldsymbol{\xi},t) := S(\boldsymbol{x},\boldsymbol{\xi},t) - \boldsymbol{x}\boldsymbol{\xi} + Et \tag{3.21}$$

in all variables $\boldsymbol{x}$, $\boldsymbol{\xi}$ and t. Condition (3.20) is thus modified to

$$\nabla_{\boldsymbol{\xi}} S(\boldsymbol{x},\boldsymbol{\xi},t) = \boldsymbol{x} \,, \tag{3.22}$$

i.e. we have to consider closed orbits in configuration space. The second condition reads

$$\nabla_{\boldsymbol{x}} S(\boldsymbol{x},\boldsymbol{\xi},t) = \boldsymbol{\xi} \,, \tag{3.23}$$

i.e. the contributing trajectories are also closed in momentum space and, hence, we are talking about periodic orbits. The third condition

$$\frac{\partial S}{\partial t}(\boldsymbol{x},\boldsymbol{\xi},t) = -E \tag{3.24}$$

finally fixes the energy of the respective periodic orbit to the value of the external parameter E. Consequently the phase at a stationary point is given by

$$\phi\Big|_{\nabla_{\boldsymbol{x}}\phi=\nabla_{\boldsymbol{\xi}}\phi=\partial\phi/\partial t=0} = \oint \boldsymbol{p}\,\mathrm{d}\boldsymbol{x} = S_{\text{p.o.}} \,, \tag{3.25}$$

the action of the orbit. Summarising, without any additional assumptions, we have already found that the spectral sum (3.18) has a semiclassical expansion of the form

$$\sum_n \varrho\left(\frac{E_n - E}{\hbar}\right) \sim \sum_{\text{p.o.}} \frac{\hat{\varrho}(T_{\text{p.o.}})}{2\pi} \mathcal{A}_{\text{p.o.}}\, \mathrm{e}^{(\mathrm{i}/\hbar)S_{\text{p.o.}}} \,, \qquad \hbar \to 0 \,, \tag{3.26}$$

where on the right hand side we have a formal sum over all classical periodic orbits with energy E and periods $T_{\text{p.o.}}$. The amplitudes $\mathcal{A}_{\text{p.o.}}$ will be specified further in the following sections.

3.3 Semiclassical Time Evolution for Pauli Hamiltonians

We now turn to the derivation of a semiclassical time evolution operator for the Pauli equation with Hamiltonian (3.2). To this end we make a WKB-like ansatz for the time evolution kernel $K(\boldsymbol{x},\boldsymbol{y},t)$. In analogy to ordinary WKB wave functions,

$$\Psi(\boldsymbol{x},t) = [a_0(\boldsymbol{x},t) + \hbar a_1(\boldsymbol{x},t) + \ldots]\, \mathrm{e}^{(\mathrm{i}/\hbar)S(\boldsymbol{x},t)} \ , \tag{3.27}$$

see e.g. [34, Chap. 6], we could make an ansatz of the form

$$K(\boldsymbol{x},\boldsymbol{y},t) = [\tilde{a}_0(\boldsymbol{x},\boldsymbol{y},t) + \hbar \tilde{a}_1(\boldsymbol{x},\boldsymbol{y},t) + \ldots]\, \mathrm{e}^{(\mathrm{i}/\hbar)\tilde{S}(\boldsymbol{x},\boldsymbol{y},t)} \tag{3.28}$$

with a scalar $\tilde{S}$, expected to turn out to be a classical action, and matrix valued amplitudes $\tilde{a}_j$. However, it is more convenient to use a Fourier representation changing from the initial position $\boldsymbol{y}$ to the conjugate variable $\boldsymbol{\xi}$ by the semiclassical ansatz

$$K(\boldsymbol{x},\boldsymbol{y},t) = \frac{1}{(2\pi\hbar)^d} \int_{\mathbb{R}^d} a_\hbar(\boldsymbol{x},\boldsymbol{\xi},t)\, \mathrm{e}^{(\mathrm{i}/\hbar)[S(\boldsymbol{x},\boldsymbol{\xi},t) - \boldsymbol{y}\boldsymbol{\xi}]}\, \mathrm{d}^d\xi \ . \tag{3.29}$$

Here S is still scalar and the matrix valued amplitude $a_\hbar$ has a semiclassical expansion of the form

$$a_\hbar(\boldsymbol{x},\boldsymbol{\xi},t) = \sum_{k\geq 0} \left(\frac{\hbar}{\mathrm{i}}\right)^k a_k(\boldsymbol{x},\boldsymbol{\xi},t) \ . \tag{3.30}$$

The initial condition (3.13) translates to

$$S(\boldsymbol{x},\boldsymbol{\xi},0) = \boldsymbol{x}\boldsymbol{\xi} \quad \text{and} \quad a_k(\boldsymbol{x},\boldsymbol{\xi},0) = \begin{cases} \mathbb{1}_{2s+1} & , \ k = 0 \\ 0 & , \ k \geq 1 \end{cases} . \tag{3.31}$$

An evaluation of (3.29) with the method of stationary phase will in general yield a superposition of several terms of the form (3.28), since in general there is more than one stationary point. We now insert the ansatz (3.29) into the Pauli equation (3.14) and sort by powers of $\hbar$. In order to simplify the presentation, here we do this for the Pauli Hamiltonian

$$\hat{H}_\mathrm{P} = -\frac{\hbar^2}{2m}\Delta + V(\boldsymbol{x}) - \frac{\hbar}{2}\frac{e}{mc}\mathrm{d}\pi_s(\boldsymbol{\sigma})\boldsymbol{B}(\boldsymbol{x}) \tag{3.32}$$

with scalar potential $V(\boldsymbol{x})$ and magnetic field $\boldsymbol{B}(\boldsymbol{x})$. The result directly generalises to the case of arbitrary matrix valued Weyl operators as we demonstrate in Appendix E. Upon inserting (3.29) into (3.14) we find the condition

$$\begin{aligned} &\left[\frac{(\nabla_{\boldsymbol{x}}S)^2}{2m} + V(\boldsymbol{x}) + \frac{\partial S}{\partial t}\right] a_0 + \hbar \left[\frac{(\nabla_{\boldsymbol{x}}S)^2}{2m} + V(\boldsymbol{x}) + \frac{\partial S}{\partial t}\right] a_1 \\ &+ \hbar \left(\frac{\partial}{\partial t} + \frac{\nabla_{\boldsymbol{x}}S}{m}\nabla_{\boldsymbol{x}} + \frac{\Delta_{\boldsymbol{x}}S}{2m} - \frac{\mathrm{i}e}{2mc}\,\mathrm{d}\pi_s(\boldsymbol{\sigma})\boldsymbol{B}(\boldsymbol{x})\right) a_0 + \mathcal{O}\left(\hbar^2\right) = 0 \ , \end{aligned} \tag{3.33}$$

where we have suppressed the arguments of S and a_j for the sake of legibility. In the case of the general Pauli Hamiltonian (3.2) one finds, see Appendix E,

$$\begin{aligned} &\left[H(\nabla_{\boldsymbol{x}} S, \boldsymbol{x}) + \frac{\partial S}{\partial t}\right] a_0 + \hbar \left[H(\nabla_{\boldsymbol{x}} S, \boldsymbol{x}) + \frac{\partial S}{\partial t}\right] a_1 \\ &+ \hbar \left(\frac{\partial}{\partial t} + \nabla_{\boldsymbol{p}} H(\nabla_{\boldsymbol{x}} S, \boldsymbol{x})\, \nabla_{\boldsymbol{x}} \right. \\ &\qquad \left. + \frac{1}{2}\, \nabla_{\boldsymbol{x}} \left(\nabla_{\boldsymbol{p}} H(\nabla_{\boldsymbol{x}} S, \boldsymbol{x})\right) + \frac{\mathrm{i}}{2}\, \mathrm{d}\pi_s(\boldsymbol{\sigma}) \boldsymbol{\mathcal{B}}(\nabla_{\boldsymbol{x}} S, \boldsymbol{x}) \right) a_0 + \mathcal{O}\left(\hbar^2\right) = 0\,, \end{aligned} \tag{3.34}$$

For derivatives of the classical Hamiltonian H we use the convention that $\nabla_{\boldsymbol{p}}$ denotes a derivative with respect to the first argument (the momentum), and $\nabla_{\boldsymbol{x}}$ is the derivative with respect to the second argument (the position). One easily verifies that (3.33) is a special case of (3.34). Solving order by order in $\hbar$, in leading order we simply find the Hamilton–Jacobi equation

$$H(\nabla_{\boldsymbol{x}} S, \boldsymbol{x}) + \frac{\partial S}{\partial t} = 0 \tag{3.35}$$

as in the case of the corresponding Schrödinger equation (i.e., as for $\boldsymbol{\mathcal{B}} \equiv 0$). From this fact and from the initial condition (3.31) Hamilton–Jacobi theory, see e.g. [35, 36], allows us to conclude that $S(\boldsymbol{x}, \boldsymbol{\xi}, t)$ generates a canonical transformation from $(\nabla_{\boldsymbol{x}} S, \boldsymbol{x})$ to $(\boldsymbol{\xi}, \nabla_{\boldsymbol{\xi}} S)$ and that these two points in phase space are connected by a trajectory of the Hamiltonian flow, i.e.

$$\phi_H^t(\boldsymbol{\xi}, \nabla_{\boldsymbol{\xi}} S) = (\nabla_{\boldsymbol{x}} S, \boldsymbol{x})\,. \tag{3.36}$$

We will denote this trajectory by $(\boldsymbol{P}(t'), \boldsymbol{X}(t'))$ solving Hamilton's equations of motion

$$\frac{\mathrm{d}\boldsymbol{X}}{\mathrm{d}t'} = \nabla_{\boldsymbol{p}} H(\boldsymbol{P}, \boldsymbol{X})\,, \quad \frac{\mathrm{d}\boldsymbol{P}}{\mathrm{d}t'} = -\nabla_{\boldsymbol{x}} H(\boldsymbol{P}, \boldsymbol{X})\,, \tag{3.37}$$

with $(\boldsymbol{P}(0), \boldsymbol{X}(0)) = (\boldsymbol{\xi}, \nabla_{\boldsymbol{\xi}} S(\boldsymbol{x}, \boldsymbol{\xi}, t))$ and $(\boldsymbol{P}(t), \boldsymbol{X}(t)) = (\nabla_{\boldsymbol{x}} S(\boldsymbol{x}, \boldsymbol{\xi}, t), \boldsymbol{x})$. The next-to-leading order equation reads

$$\begin{aligned} &\left(H(\nabla_{\boldsymbol{x}} S, \boldsymbol{x}) + \frac{\partial S}{\partial t}\right) a_1(\boldsymbol{x}, \boldsymbol{\xi}, t) + \left[\nabla_{\boldsymbol{p}} H(\nabla_{\boldsymbol{x}} S, \boldsymbol{x}) \nabla_{\boldsymbol{x}} + \frac{\partial}{\partial t} \right. \\ &\left. + \frac{1}{2} \nabla_{\boldsymbol{x}} \left(\nabla_{\boldsymbol{p}} H(\nabla_{\boldsymbol{x}} S, \boldsymbol{x})\right) + \frac{\mathrm{i}}{2} \boldsymbol{\mathcal{B}}(\nabla_{\boldsymbol{x}} S, \boldsymbol{x})\, \mathrm{d}\pi_s(\boldsymbol{\sigma}) \right] a_0(\boldsymbol{x}, \boldsymbol{\xi}, t) = 0\,. \end{aligned} \tag{3.38}$$

Due to the $\hbar^0$-equation the Hamilton–Jacobi equation is solved by $S(\boldsymbol{x}, \boldsymbol{\xi}, t)$ and thus the term in front of $a_1(\boldsymbol{x}, \boldsymbol{\xi}, t)$ drops out of (3.38). Realising that the derivative terms in the square bracket constitute a total time derivative along the trajectory generated by $S(\boldsymbol{x}, \boldsymbol{\xi}, t)$ we define

$$\frac{\mathrm{d}}{\mathrm{d}t} := \frac{\partial}{\partial t} + \left(\nabla_{\boldsymbol{p}} H(\nabla_{\boldsymbol{x}} S, \boldsymbol{x})\right) \nabla_{\boldsymbol{x}}\,. \tag{3.39}$$

Therefore, we are left with an equation for $a_0(\boldsymbol{x},\boldsymbol{\xi},t)$ reading

$$\left[\frac{\mathrm{d}}{\mathrm{d}t}+\frac{1}{2}\nabla_{\boldsymbol{x}}\left(\nabla_{\boldsymbol{p}}H(\nabla_{\boldsymbol{x}}S,\boldsymbol{x})\right)+\frac{\mathrm{i}}{2}\boldsymbol{\mathcal{B}}(\nabla_{\boldsymbol{x}}S,\boldsymbol{x})\,\mathrm{d}\pi_s(\boldsymbol{\sigma})\right]a_0(\boldsymbol{x},\boldsymbol{\xi},t)\,. \tag{3.40}$$

The solution of the scalar transport equation – i.e. (3.40) with $\boldsymbol{\mathcal{B}}\equiv 0$ – is known to be given by $\sqrt{\det(\partial^2 S/\partial x\partial\xi)}$ (we give a derivation of this result in Appendix B). Thus, we can make the ansatz

$$a_0(\boldsymbol{x},\boldsymbol{\xi},t)=\sqrt{\det\frac{\partial^2 S}{\partial x\partial\xi}(\boldsymbol{x},\boldsymbol{\xi},t)}\;D(\boldsymbol{x},\boldsymbol{\xi},t) \tag{3.41}$$

with a $(2s+1)\times(2s+1)$ matrix $D(\boldsymbol{x},\boldsymbol{\xi},t)$, which is still to be determined. Upon inserting (3.41) into (3.40) we retain the spin transport equation

$$\dot{D}(\boldsymbol{x},\boldsymbol{\xi},t)+\frac{\mathrm{i}}{2}\boldsymbol{\mathcal{B}}(\nabla_{\boldsymbol{x}}S,\boldsymbol{x})\,\mathrm{d}\pi_s(\boldsymbol{\sigma})\,D(\boldsymbol{x},\boldsymbol{\xi},t)=0 \tag{3.42}$$

with initial condition $D(\boldsymbol{x},\boldsymbol{\xi},0)=\mathbb{1}_{2s+1}$, where the dot denotes the time derivative $\mathrm{d}/\mathrm{d}t$. Up to now the notation (3.39) was a mere abbreviation, but if we rewrite (3.42) in terms of the trajectory $(\boldsymbol{P}(t'),\boldsymbol{X}(t'))$, we get a linear ordinary differential equation along a flow line of ϕ_H^t, cf. [37],

$$\dot{D}(\boldsymbol{\xi},\boldsymbol{y},t)+\frac{\mathrm{i}}{2}\boldsymbol{\mathcal{B}}(\boldsymbol{P}(t),\boldsymbol{X}(t))\,\mathrm{d}\pi_s(\boldsymbol{\sigma})\,D(\boldsymbol{\xi},\boldsymbol{y},t)=0\,, \tag{3.43}$$

with initial condition $D(\boldsymbol{\xi},\boldsymbol{x},0)=\mathbb{1}_{2s+1}$. Notice that for D on the trajectory $(\boldsymbol{P}(t),\boldsymbol{X}(t))$ we have introduced a new notation, since the solution of (3.43) indeed depends only on the initial point $(\boldsymbol{\xi},\boldsymbol{y})$ in phase space at which we start with the integration and on the time t until which we proceed. In the following we will always use this new notation; thus, it cannot be confused with the old convention which was exclusively used in (3.41) and (3.42).

The spin transport equation (3.43) has the unique solution

$$D(\boldsymbol{\xi},\boldsymbol{y},t)=\left[\exp\left(-\frac{\mathrm{i}}{2}\int_0^t\boldsymbol{\mathcal{B}}(\boldsymbol{P}(t'),\boldsymbol{X}(t'))\,\mathrm{d}\pi_s(\boldsymbol{\sigma})\,\mathrm{d}t'\right)\right]_+ \tag{3.44}$$

where $[\ldots]_+$ denotes a time ordered expression, i.e. terms with smallest time arguments are moved to the very right. Since the exponent in (3.44) is an element of a derived representation of the Lie algebra $\mathfrak{su}(2)$ the solution $d(\boldsymbol{x},\boldsymbol{\xi},t)$ itself is the representation of some element of the group SU(2). One immediately sees that, equivalently, we may formulate the spin transport equation itself as an equation on the group,

$$\dot{d}(\boldsymbol{\xi},\boldsymbol{y},t)+\frac{\mathrm{i}}{2}\boldsymbol{\mathcal{B}}(\boldsymbol{P}(t),\boldsymbol{X}(t))\,\boldsymbol{\sigma}\,d(\boldsymbol{\xi},\boldsymbol{y},t)=0\,,\quad d(\boldsymbol{\xi},\boldsymbol{y},0)=id\,, \tag{3.45}$$

recovering D by

$$D(\boldsymbol{\xi}, \boldsymbol{y}, 0) = \pi_s(d(\boldsymbol{\xi}, \boldsymbol{y}, 0)) \,. \tag{3.46}$$

Thus, whatever spin the particle described has, we obtain the solution of the transport equation by solving (3.45) and then choosing the appropriate representation.

We could now insert the results obtained so far into ansatz (3.29) and evaluate the $\boldsymbol{\xi}$-integration with the method of stationary phase in order to obtain a semiclassical time evolution kernel for the Pauli equation. Since in the following we will not rely on this semiclassical kernel but rather directly employ the intermediate results, we refrain from doing this here but refer the reader to [1–3] where this issue is dealt with in detail.

3.4 Spin Transport and Spin Precession

We now want to discuss some properties of the spin transport equation

$$\dot{d}(\boldsymbol{p}, \boldsymbol{x}, t) + \frac{\mathrm{i}}{2} \boldsymbol{\sigma} \boldsymbol{\mathcal{B}} \, d(\boldsymbol{p}, \boldsymbol{x}, t) = 0 \,, \tag{3.47}$$

which we will heavily make use of throughout the rest of the work. In the previous section we have already seen that the existence and uniqueness theorem for ordinary differential equations guarantees a unique solution of (3.47) (at least for continuous fields $\boldsymbol{\mathcal{B}}$) with initial condition $d(\boldsymbol{p}, \boldsymbol{x}, 0) = id$. From the same theorem we can also conclude the composition law

$$d(\boldsymbol{p}, \boldsymbol{x}, t + t') = d(\phi_H^t(\boldsymbol{p}, \boldsymbol{x}), t') \, d(\boldsymbol{p}, \boldsymbol{x}, t) \,, \tag{3.48}$$

which makes d an SU(2)-valued cocycle of the flow ϕ_H^t.

Now consider the situation where $(\boldsymbol{p}, \boldsymbol{x})$ and $(\boldsymbol{p}', \boldsymbol{x}')$ are both points on the same periodic orbit γ with period T_γ, i.e. we have $(\boldsymbol{p}', \boldsymbol{x}') = \phi_H^t(\boldsymbol{p}, \boldsymbol{x})$ for some $0 \le t \le T_\gamma$ and, equivalently, $(\boldsymbol{p}, \boldsymbol{x}) = \phi_H^{-t}(\boldsymbol{p}', \boldsymbol{x}') = \phi_H^{T_\gamma - t}(\boldsymbol{p}', \boldsymbol{x}')$. Let us investigate how $d(\boldsymbol{p}, \boldsymbol{x}, T_\gamma)$ and $d(\boldsymbol{p}', \boldsymbol{x}', T_\gamma)$ are related. From (3.48) it follows that

$$\begin{aligned} d(\boldsymbol{p}', \boldsymbol{x}', T_\gamma) &= d(\phi_H^{T_\gamma - t}(\boldsymbol{p}', \boldsymbol{x}'), t) \, d(\phi_H^{-t}(\boldsymbol{p}', \boldsymbol{x}'), T_\gamma) \, d(\boldsymbol{p}', \boldsymbol{x}', -t) \\ &= d(\boldsymbol{p}, \boldsymbol{x}, t) \, d(\boldsymbol{p}, \boldsymbol{x}, T_\gamma) \, d(\boldsymbol{p}', \boldsymbol{x}', -t) \,. \end{aligned} \tag{3.49}$$

Since, in addition we know that

$$\begin{aligned} id = d(\boldsymbol{p}', \boldsymbol{x}', 0) &= d(\phi^{-t}(\boldsymbol{p}', \boldsymbol{x}'), t) \, d(\boldsymbol{p}', \boldsymbol{x}', -t) \\ &= d(\boldsymbol{p}, \boldsymbol{x}, t) \, d(\boldsymbol{p}', \boldsymbol{x}', -t) \,, \end{aligned} \tag{3.50}$$

we see that $d(\boldsymbol{p}, \boldsymbol{x}, T_\gamma)$ and $d(\boldsymbol{p}', \boldsymbol{x}', T_\gamma)$ are conjugate to each other,

$$d(\boldsymbol{p}', \boldsymbol{x}', T_\gamma) = d(\boldsymbol{p}, \boldsymbol{x}, t) d(\boldsymbol{p}, \boldsymbol{x}, T_\gamma) [d(\boldsymbol{p}, \boldsymbol{x}, t)]^{-1} \,. \tag{3.51}$$

This is in so far an important observation as in trace formulae we need the matrix trace of $D(\boldsymbol{p},\boldsymbol{x},t) = \pi_s(d(\boldsymbol{p},\boldsymbol{x},t))$, $(\boldsymbol{p},\boldsymbol{x}) \in \gamma$, and (3.51) guarantees that this object is independent of the point $(\boldsymbol{p},\boldsymbol{x}) \in \gamma$ where we start the integration of (3.47). Thus we may define (with $(\boldsymbol{p},\boldsymbol{x}) \in \gamma$ arbitrary)

$$\mathrm{tr}\, D_\gamma := \mathrm{tr}\, \pi_s(d(\boldsymbol{p},\boldsymbol{x},T_\gamma)) . \tag{3.52}$$

For the spin transport we have to solve an equation in SU(2) along the flow lines of ϕ_H^t. Thus we are dealing with a mixed quantum/classical representation: The translational degrees of freedom are purely classical, whereas the spin transporter D is the time evolution operator for the propagation of a $(2s+1)$-component spinor along a classical trajectory. We will now show how to reinterpret this situation in a purely classical context where the spin degree of freedom is described by a classical spin vector $\boldsymbol{s}$ of fixed length, i.e. $\boldsymbol{s} \in S^2 \hookrightarrow \mathbb{R}^3$. To this end recall that the adjoint representation of SU(2),

$$\begin{aligned} \mathrm{Ad}_g : \quad \mathfrak{su}(2) &\rightarrow \mathfrak{su}(2) \\ Z = \boldsymbol{\sigma z} &\mapsto gZg^{-1} = \boldsymbol{\sigma}(\varphi(g)\boldsymbol{z}) \end{aligned} \tag{3.53}$$

yields the double covering map from SU(2) to SO(3),

$$\varphi : g \mapsto \varphi(g) \in \mathrm{SO}(3) , \tag{3.54}$$

cf. Appendix C. In this way we can associate with every $d(\boldsymbol{p},\boldsymbol{x},t) \in \mathrm{SU}(2)$ a rotation

$$R(\boldsymbol{p},\boldsymbol{x},t) := \varphi(d(\boldsymbol{p},\boldsymbol{x},t)) . \tag{3.55}$$

Let us determine the equation of motion which $R(\boldsymbol{p},\boldsymbol{x},t)$ fulfills:

$$\begin{aligned} \frac{\mathrm{d}}{\mathrm{d}t}\boldsymbol{\sigma}(R(\boldsymbol{p},\boldsymbol{x},t)\boldsymbol{z}) &= \frac{\mathrm{d}}{\mathrm{d}t}[d(\boldsymbol{p},\boldsymbol{x},t)(\boldsymbol{\sigma z})d^{-1}(\boldsymbol{p},\boldsymbol{x},t)] \\ &= \dot{d}(\boldsymbol{z\sigma})d^{-1} + d(\boldsymbol{\sigma z})\dot{d}^{-1} \\ &\underset{(*)}{=} -\frac{\mathrm{i}}{2}(\boldsymbol{\sigma}\boldsymbol{\mathcal{B}})d(\boldsymbol{\sigma z})d^{-1} + d(\boldsymbol{\sigma z})d^{-1}\frac{\mathrm{i}}{2}(\boldsymbol{\sigma}\boldsymbol{\mathcal{B}}) \\ &= \frac{\mathrm{i}}{2}[\boldsymbol{\sigma}(R\boldsymbol{z}), \boldsymbol{\sigma}\boldsymbol{\mathcal{B}}] \\ &= \boldsymbol{\sigma}(\boldsymbol{\mathcal{B}} \times (R\boldsymbol{z})) , \end{aligned} \tag{3.56}$$

where in $(*)$ we have used (3.47) and the fact that d is unitary, i.e.

$$\dot{d}^{-1} = \dot{d}^\dagger = d^\dagger \left(\tfrac{\mathrm{i}}{2}\boldsymbol{\sigma}\boldsymbol{\mathcal{B}}\right) . \tag{3.57}$$

Writing the vector product as a matrix multiplication in $\mathbb{R}^{3,3}$ we obtain

$$\frac{\mathrm{d}}{\mathrm{d}t}\boldsymbol{\sigma}(R(\boldsymbol{p},\boldsymbol{x},t)\boldsymbol{z}) = \boldsymbol{\sigma}(A_{\boldsymbol{\mathcal{B}}}(R(\boldsymbol{p},\boldsymbol{x},t)\boldsymbol{z})) \tag{3.58}$$

with

$$A_{\mathcal{B}} := \begin{pmatrix} 0 & -\mathcal{B}_z & \mathcal{B}_y \\ \mathcal{B}_z & 0 & -\mathcal{B}_x \\ -\mathcal{B}_y & \mathcal{B}_x & 0 \end{pmatrix} . \tag{3.59}$$

Thus the equation of motion for $R(\boldsymbol{p}, \boldsymbol{x}, t)$ finally reads

$$\dot{R}(\boldsymbol{p}, \boldsymbol{x}, t) = A_{\mathcal{B}(\phi^t_H(\boldsymbol{p},\boldsymbol{x}))}\, R(\boldsymbol{p}, \boldsymbol{x}, t) . \tag{3.60}$$

Now notice that the solution of

$$\dot{\boldsymbol{s}} = A_{\mathcal{B}}\, \boldsymbol{s} = \mathcal{B} \times \boldsymbol{s} , \quad \boldsymbol{s}\big|_{t=0} = \boldsymbol{s}_0 \in S^2 \hookrightarrow \mathbb{R}^3 , \tag{3.61}$$

is given by $R(\boldsymbol{p}, \boldsymbol{x}, t)\boldsymbol{s}_0$, i.e. the covering map indeed allows us to calculate the time evolution operator for a classical spin vector $\boldsymbol{s}$ from the time evolution operator $d(\boldsymbol{p}, \boldsymbol{x}, t)$ of a quantum mechanical spinor. Moreover, this calculation shows that the classical equation of spin precession (3.61) naturally emerges in a semiclassical analysis of the respective quantum problem, namely the Pauli equation.

Since $\varphi : \mathrm{SU}(2) \to \mathrm{SO}(3)$ is a double covering we certainly loose information when passing from $d(\boldsymbol{p}, \boldsymbol{x}, t)$ to $R(\boldsymbol{p}, \boldsymbol{x}, t)$. Thus it remains to show that in semiclassical trace formulae – the ultimate result we are aiming at in this chapter – everything can be conveniently expressed in terms of the classical spin. To this end recall that in trace formulae we are not interested in the full $d(\boldsymbol{p}, \boldsymbol{x}, t)$ but only in $\mathrm{tr}\, D(\boldsymbol{p}, \boldsymbol{x}, t) = \mathrm{tr}\, \pi_s(d(\boldsymbol{p}, \boldsymbol{x}, t))$, i.e. in the character χ_s of the $(2s+1)$-dimensional representation of SU(2). Parameterising SU(2) as,

$$\mathrm{SU}(2) \ni g = \mathrm{e}^{(\mathrm{i}/2)\alpha\boldsymbol{\sigma}\boldsymbol{n}} , \quad \alpha \in [0, 4\pi) , \quad \boldsymbol{n} \in S^2 \hookrightarrow \mathbb{R}^3 , \tag{3.62}$$

it is well-known that χ_s depends only on the angle α but not on the axis $\boldsymbol{n}$. Explicitly we have, see Appendix C,

$$\chi_s(\alpha) = \sum_{m=-s}^{s} \mathrm{e}^{im\alpha} . \tag{3.63}$$

On the other hand, one knows, see e.g. Appendix C, that $\varphi(\exp(\mathrm{i}\alpha\boldsymbol{\sigma}\boldsymbol{n}/2))$ is a rotation about an axis $\boldsymbol{n}$ by an angle α. Thus, the weight factor $\chi_s(\alpha_\gamma) := \mathrm{tr}\, D_\gamma$ entering the semiclassical trace formula can be fully interpreted in terms of classical spin precession as follows: Propagating an arbitrary spin vector $\boldsymbol{s}_0$ along a periodic orbit γ with the classical equation of spin precession (3.61), it is rotated by $R(\boldsymbol{p}, \boldsymbol{x}, t)$ as illustrated in Fig. 3.1. Only the rotation angle α_γ, which, in contrast to the axis $\boldsymbol{n}$, is invariant under a shift along the orbit, enters the trace formula.

Concerning the parameterisation (3.62) we remark that in [1–3] instead of using a rotation axis $\boldsymbol{n}$ and angle ϕ the respective objects where parameterised in terms of Euler angles (θ, ϕ, η). The relation with the expressions in the present work is then given by

$$2\cos(\alpha/2) = 2\cos(\theta/2)\cos(\eta) . \tag{3.64}$$

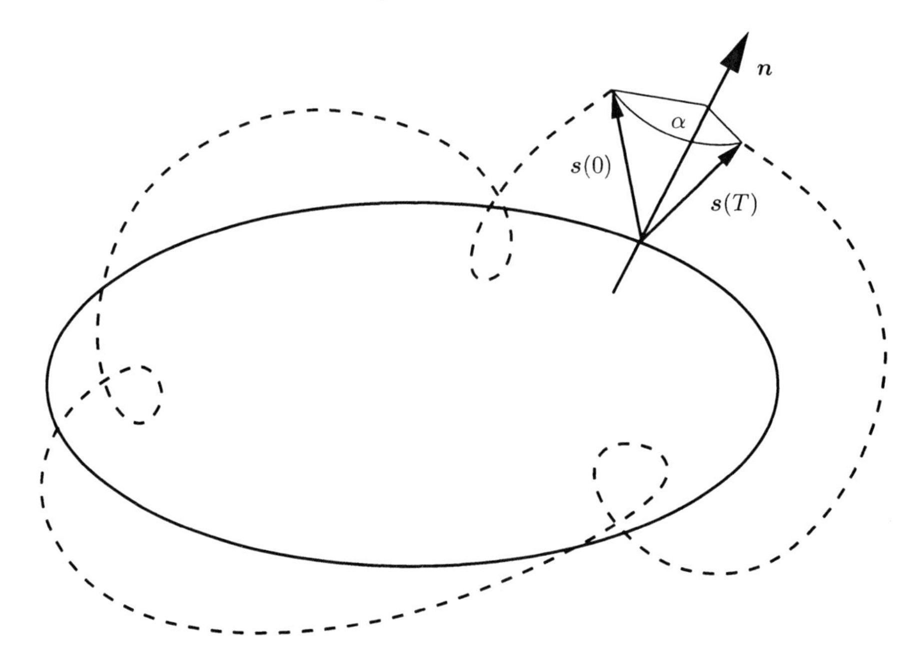

Fig. 3.1. Schematic illustration of an initial spin vector $\boldsymbol{s}(0)$ which is propagated along a periodic obit (*solid line*) according to the equation of classical spin precession (3.61). After one cycle $\boldsymbol{s}(0)$ has been rotated by an angle α about the axis $\boldsymbol{n}$, i.e. $\boldsymbol{s}(T) = \varphi(\exp(\mathrm{i}\alpha\boldsymbol{\sigma n})/2)\boldsymbol{s}(0)$.

Finally, we remark that in the classical picture we still find a remnant of the peculiar behaviour of spinors for half-integer spin, which have to be rotated by 4π before they return to the initial state. Namely, the weight factor for a rotation by 2π is given by

$$\chi_s(2\pi) = \sum_{m=-s}^{s} \mathrm{e}^{2\pi\mathrm{i}m} = (-1)^{2s}\,(2s+1)\,, \tag{3.65}$$

i.e., for half integer s we find a minus sign.

3.5 Semiclassical Trace Formulae

We have seen in Sect. 3.2 that in order to find a semiclassical trace formula we have to analyse the expression

$$\sum_n \varrho\left(\frac{E_n - E}{\hbar}\right)$$
$$= \int_{\mathbb{R}^d}\int_{\mathbb{R}}\int_{\mathbb{R}^d} \frac{\hat{\varrho}(t)}{2\pi}\,\frac{\operatorname{tr} a_0(\boldsymbol{x},\boldsymbol{\xi},t)}{(2\pi\hbar)^d}\,\mathrm{e}^{(\mathrm{i}/\hbar)(S(\boldsymbol{x},\boldsymbol{\xi},t)-\boldsymbol{x}\boldsymbol{\xi}+Et)}\,\mathrm{d}^d\xi\,\mathrm{d}t\,\mathrm{d}^d x\,[1+\mathcal{O}(\hbar)]\,, \tag{3.66}$$

where tr denotes the matrix trace on $\mathbb{C}^{2s+1}$. The integrals shall be evaluated by the method of stationary phase and we will take advantage of our knowledge of $S(\boldsymbol{x},\boldsymbol{\xi},t)$ and $a_0(\boldsymbol{x},\boldsymbol{\xi},t)$ gained so far. In section 3.2 we already noticed that the stationary points are given by periodic orbits of energy E of the flow ϕ_H^t with classical Hamiltonian $H(\boldsymbol{p},\boldsymbol{x})$. Obviously the stationary points are never isolated but appear in families. Even an isolated periodic orbit gives rise to a one-parameter family of stationary points. In general we find submanifolds $\mathcal{M}_s \subset \mathbb{R}^d \times \mathbb{R}^d \times \mathbb{R}$ of stationary points with dimension $d_s := \dim \mathcal{M}_s$. Therefore, we have $2d+1-d_s$ integrals which can be evaluated by the method of stationary phase, and the remaining d_s-dimensional integral over $\mathcal{M}_s$ has to be performed exactly, yielding the general trace formula

$$\sum_n \varrho\left(\frac{E_n - E}{\hbar}\right) = \sum_{\mathcal{M}_s} \frac{\hat{\varrho}(T_{\mathcal{M}_s})}{2\pi}\mathcal{A}_{\mathcal{M}_s}\,\mathrm{e}^{(\mathrm{i}/\hbar)S_{\mathcal{M}_s}}(1+\mathcal{O}(\hbar))\,,\quad \hbar\to 0\,, \tag{3.67}$$

where $T_{\mathcal{M}_s}$ and $S_{\mathcal{M}_s}$ denote the period and the action, respectively, of the periodic orbits in $\mathcal{M}_s$. From (3.66) and from Theorem 3 (see appendix D) we also conclude that the semiclassical order of a contribution coming from $\mathcal{M}_s$ is given by

$$\mathcal{A}_{\mathcal{M}_s} = \mathcal{O}\left(\hbar^{\frac{1-d_s}{2}}\right)\,. \tag{3.68}$$

We will explicitly treat the following three cases:

(i) The energy shell

$$\Omega_E := \{(\boldsymbol{p},\boldsymbol{x})\,|\,H(\boldsymbol{p},\boldsymbol{x}) = E\} \tag{3.69}$$

is a set of periodic orbits of period $T=0$, i.e. we have to consider

$$\mathcal{M}_s = \Omega_E \times \{0\} \subset \mathbb{R}^d \times \mathbb{R}^d \times \mathbb{R} \tag{3.70}$$

with $d_s = 2d-1$.

(ii) An isolated periodic orbit γ together with its period T_γ constitutes a set of stationary points with dimension $d_s = 1$.

(iii) Rational Liouville–Arnold tori, which appear in integrable systems, see also Sect. 4.4.1, together with their period $T_{\boldsymbol{m}}$ form a set of stationary points of dimension d.

In principal, the contribution of any family of periodic orbits can be calculated; in [38] a general method was developed for the typical case that $\mathcal{M}_s$, with $d_s > 1$, is generated by the action of a group on one orbit $\gamma \in \mathcal{M}_s$.

We remark that a trace formula for the density of states can, formally, be obtained by choosing $\varrho(\omega) = \delta(\hbar\omega)$, yielding

$$d(E) \sim \sum_{\mathcal{M}_s} \frac{\mathcal{A}_{\mathcal{M}_s}}{2\pi\hbar} \, \mathrm{e}^{(\mathrm{i}/\hbar) S_{\mathcal{M}_s}} (1 + \mathcal{O}(\hbar)) \, , \quad \hbar \to 0 \, , \tag{3.71}$$

which has to be read as a distributional identity meaning (3.67).

3.5.1 The Weyl term

We start with the calculation of the contribution which comes from $t = 0$. Since for $t = 0$ the phase in (3.66) vanishes, this term is not rapidly oscillating but instead constitutes the mean behaviour of the spectral sum in (3.66). The contribution is known as the Thomas–Fermi or Weyl term.

Expanding the total phase

$$\phi(\boldsymbol{x}, \boldsymbol{\xi}, t) := S(\boldsymbol{x}, \boldsymbol{\xi}, t) - \boldsymbol{x}\boldsymbol{\xi} + Et \tag{3.72}$$

about $t = 0$, we find

$$\begin{aligned}
\phi(\boldsymbol{x}, \boldsymbol{\xi}, t) &= \left(\frac{\partial S}{\partial t}(\boldsymbol{x}, \boldsymbol{\xi}, t) + E \right) t + \frac{1}{2} \frac{\partial^2 S}{\partial t^2}(\boldsymbol{x}, \boldsymbol{\xi}, t) \, t^2 + \mathcal{O}\left(t^3\right) \\
&= (E - H(\boldsymbol{\xi}, \boldsymbol{x})) \, t - \frac{1}{2} (\nabla_{\boldsymbol{p}} H(\boldsymbol{\xi}, \boldsymbol{x})) \left[\frac{\partial}{\partial t} \nabla_{\boldsymbol{x}} S(\boldsymbol{x}, \boldsymbol{\xi}, t) \right]_{t=0} t^2 + \mathcal{O}\left(t^3\right) \\
&= (E - H(\boldsymbol{\xi}, \boldsymbol{x})) \, t + \frac{1}{2} (\nabla_{\boldsymbol{p}} H(\boldsymbol{\xi}, \boldsymbol{x}))(\nabla_{\boldsymbol{x}} H(\boldsymbol{\xi}, \boldsymbol{x})) \, t^2 + \mathcal{O}\left(t^3\right) \, ,
\end{aligned} \tag{3.73}$$

where we have used a couple of times that at stationary points $S(\boldsymbol{x}, \boldsymbol{\xi}, t)$ solves the Hamilton–Jacobi equation (3.35). Introducing polar coordinates in the $\boldsymbol{\xi}$-integral by $\boldsymbol{\xi} = \xi\boldsymbol{\omega}$, $\xi \in [0, \infty)$, $\boldsymbol{\omega} \in S^{d-1} \hookrightarrow \mathbb{R}^d$ (i.e. we write $\boldsymbol{\omega} \in S^{d-1}$ as a unit vector in $\mathbb{R}^d$), we can use the method of stationary phase in the integrals over ξ and t and perform the remaining integrations exactly. The stationary points $(t = 0, \xi = \xi_s)$ are obtained by solving

$$\frac{\partial}{\partial t} \left([E - H(\xi\boldsymbol{\omega}, \boldsymbol{x})] \, t + \mathcal{O}(t^2) \right) \Big|_{t=0} = E - H(\xi\boldsymbol{\omega}, \boldsymbol{x}) \stackrel{!}{=} 0 \tag{3.74}$$

for ξ_s. In addition we need the following matrix of second derivatives,

$$D^2\phi := \left. \begin{pmatrix} \frac{\partial^2 \phi}{\partial t^2} & \frac{\partial^2 \phi}{\partial t \partial \xi} \\ \frac{\partial^2 \phi}{\partial t \partial \xi} & \frac{\partial^2 \phi}{\partial \xi^2} \end{pmatrix} \right|_{t=0, \xi=\xi_s} \, , \tag{3.75}$$

which, with

$$\left.\frac{\partial^2\phi}{\partial t\partial\xi}\right|_{t=0,\xi=\xi_s} = -\frac{\partial}{\partial\xi}H(\xi\boldsymbol{\omega},\boldsymbol{x})\Big|_{t=0,\xi=\xi_s} = (\nabla_{\boldsymbol{p}}H(\xi_s\boldsymbol{\omega},\boldsymbol{x}))\,(-\boldsymbol{\omega}) \quad \text{and}$$
$$\left.\frac{\partial^2\phi}{\partial\xi^2}\right|_{t=0,\xi=\xi_s} = 0\,. \tag{3.76}$$

reads

$$D^2\phi = \begin{pmatrix} (\nabla_{\boldsymbol{p}}H(\boldsymbol{\xi},\boldsymbol{x}))(\nabla_{\boldsymbol{x}}H(\boldsymbol{\xi},\boldsymbol{x})) & (\nabla_{\boldsymbol{p}}H(\xi_s\boldsymbol{\omega},\boldsymbol{x}))\,(-\boldsymbol{\omega}) \\ (\nabla_{\boldsymbol{p}}H(\xi_s\boldsymbol{\omega},\boldsymbol{x}))\,(-\boldsymbol{\omega}) & 0 \end{pmatrix}. \tag{3.77}$$

The determinant of this matrix is given by

$$\det D^2\phi = \left[(\nabla_{\boldsymbol{p}}H(\xi_s\boldsymbol{\omega},\boldsymbol{x}))\,\boldsymbol{\omega}\right]^2 \tag{3.78}$$

and the signature vanishes if only $\nabla_{\boldsymbol{p}}H \neq 0$. Thus, by an application of Theorem 3 (see appendix D) the Weyl term finally becomes

$$\begin{aligned} \mathcal{A}_{\Omega_E} &= \frac{1}{(2\pi\hbar)^{d-1}}\int_{\mathbb{R}^d}\int_{S^{d-1}}\int_0^\infty \frac{\operatorname{tr} a_0(\boldsymbol{x},\xi_s\boldsymbol{\omega},0)\,\delta(\xi-\xi_s)}{|(\nabla_{\boldsymbol{p}}H(\xi_s\boldsymbol{\omega},\boldsymbol{x}))\boldsymbol{\omega}|} \\ &\qquad\qquad \times \xi^{d-1}\,\mathrm{d}\xi\,\mathrm{d}^{d-1}\omega\,\mathrm{d}^d x\,[1+\mathcal{O}(\hbar)] \\ &= \frac{2s+1}{(2\pi\hbar)^{d-1}}\int_{\mathbb{R}^d}\int_{\mathbb{R}^d}\delta(H(\boldsymbol{\xi},\boldsymbol{x})-E)\,\mathrm{d}^d\xi\,\mathrm{d}^d x\,[1+\mathcal{O}(\hbar)] \\ &= \frac{(2s+1)\,|\Omega_E|}{(2\pi\hbar)^{d-1}}\,[1+\mathcal{O}(\hbar)]\,. \end{aligned} \tag{3.79}$$

For the density of states, cf. (3.71), this term constitutes the average behaviour (since, due to the vanishing action, it does not oscillate),

$$\bar{d}(E) = \frac{(2s+1)\,|\Omega_E|}{(2\pi\hbar)^d}\,. \tag{3.80}$$

This is the familiar "one state per Planck–cell" rule, i.e. the mean behaviour of the spectral density is governed by the ratio of the phase space volume, which is accessible classically and the volume of a Planck–cell, $(2\pi\hbar)^d$, here multiplied by $(2s+1)$, the number of possible spin projections.

3.5.2 Hyperbolic Systems

The contribution to (3.67) of an isolated periodic orbit γ with period T_γ can be calculated as follows. Locally we can introduce new coordinates $(P_u, \boldsymbol{P}_v, X_u, \boldsymbol{X}_v)$ by a canonical transformation such that X_u parameterises the orbit in position space and the $\boldsymbol{X}_v$ are transversal to the orbit. The variables P_u and $\boldsymbol{P}_v$ are the corresponding canonically conjugate momenta. The flow ϕ_H^t at time $t = T_\gamma$ then maps a point near the orbit as,

$$\begin{pmatrix} P_u \\ X_u \\ \boldsymbol{P}_v \\ \boldsymbol{X}_v \end{pmatrix} \mapsto \begin{pmatrix} \mathbb{1}_2 & 0 \\ 0 & \mathbb{M}_\gamma \end{pmatrix} \begin{pmatrix} P_u \\ X_u \\ \boldsymbol{P}_v \\ \boldsymbol{X}_v \end{pmatrix} + \mathcal{O}\left(P_u^2 + X_u^2 + \boldsymbol{P}_v^2 + \boldsymbol{X}_v^2\right) , \tag{3.81}$$

where the $(2d-2) \times (2d-2)$ matrix $\mathbb{M}_\gamma$ is the monodromy matrix, the linearised Poincaré map of the orbit. Thus in a vicinity of γ the dynamics separates, i.e. we can write

$$H(\boldsymbol{P}, \boldsymbol{X}) = H_u(P_u, X_u) + H_v(\boldsymbol{P}_v, \boldsymbol{X}_v) , \tag{3.82}$$

and, accordingly, the action is also additive,

$$S(\boldsymbol{x}, \boldsymbol{\xi}, t) = S_u(x_u, \xi_u, t) + S_v(\boldsymbol{x}_v, \boldsymbol{\xi}_v, t) . \tag{3.83}$$

In (3.66) we now perform all integrals except for the x_u-integral by the method of stationary phase. To this end we need the determinant of the following matrix of second derivatives of the total phase $\phi(\boldsymbol{x}, \boldsymbol{\xi}, t) := S(\boldsymbol{x}, \boldsymbol{\xi}, t) - \boldsymbol{\xi}\boldsymbol{x} + Et$,

$$D^2\phi := \begin{pmatrix} \frac{\partial^2 S}{\partial x_v \partial x_v} & \frac{\partial^2 S}{\partial x_v \partial \xi_v} - \mathbb{1}_{d-1} & \frac{\partial^2 S}{\partial x_v \partial \xi_u} & \frac{\partial^2 S}{\partial x_v \partial t} \\ \frac{\partial^2 S}{\partial \xi_v \partial x_v} - \mathbb{1}_{d-1} & \frac{\partial^2 S}{\partial \xi_v \partial \xi_v} & \frac{\partial^2 S}{\partial \xi_v \partial \xi_u} & \frac{\partial^2 S}{\partial \xi_v \partial t} \\ \frac{\partial^2 S}{\partial \xi_u \partial x_v} & \frac{\partial^2 S}{\partial \xi_u \partial \xi_v} & \frac{\partial^2 S}{\partial \xi_u \partial \xi_u} & \frac{\partial^2 S}{\partial \xi_u \partial t} \\ \frac{\partial^2 S}{\partial t \partial x_v} & \frac{\partial^2 S}{\partial t \partial \xi_v} & \frac{\partial^2 S}{\partial t \partial \xi_u} & \frac{\partial^2 S}{\partial t \partial t} \end{pmatrix} , \tag{3.84}$$

where all terms have to be evaluated on γ, i.e. $\boldsymbol{x}_v = 0 = \boldsymbol{\xi}_v$ and $t = T_\gamma$. Due to the separation in u- and v-directions we have

$$\frac{\partial^2 S}{\partial x_v \partial \xi_u} = 0 = \frac{\partial^2 S}{\partial \xi_v \partial \xi_u} . \tag{3.85}$$

Similarly, using the Hamilton–Jacobi equation (3.35) we also find

$$\frac{\partial^2 S}{\partial x_v \partial t} = -\frac{\partial}{\partial x_v} H_v(\boldsymbol{\xi}_v, \frac{\partial S_v}{\partial \xi_v}) = -\frac{\partial H}{\partial x_v}\frac{\partial^2 S_v}{\partial \xi_v x_v} = \dot{\boldsymbol{P}}_v \frac{\partial^2 S_v}{\partial \xi_v x_v} = 0 , \tag{3.86}$$

since on the orbit $\dot{\boldsymbol{P}}_v \equiv 0$. Analogously, we have

$$\frac{\partial^2 S}{\partial \xi_v \partial t} = -\frac{\partial}{\partial \xi_v} H_v(\frac{\partial S_v}{\partial x_v}, \boldsymbol{x}_v) = -\frac{\partial H}{\partial \xi_v}\frac{\partial^2 S_v}{\partial \xi_v \xi_v} = -\dot{\boldsymbol{X}}_v \frac{\partial^2 S_v}{\partial \xi_v \xi_v} = 0 . \tag{3.87}$$

Thus (3.84) is block diagonal and the determinant of $D^2\phi$ simplifies to

$$\det D^2\phi = \det \begin{pmatrix} \frac{\partial^2 S}{\partial x_v \partial x_v} & \frac{\partial^2 S}{\partial x_v \partial \xi_v} - \mathbb{1}_{d-1} \\ \frac{\partial^2 S}{\partial \xi_v \partial x_v} - \mathbb{1}_{d-1} & \frac{\partial^2 S}{\partial \xi_v \partial \xi_v} \end{pmatrix} \det \begin{pmatrix} \frac{\partial^2 S}{\partial \xi_u \partial \xi_u} & \frac{\partial^2 S}{\partial \xi_u \partial t} \\ \frac{\partial^2 S}{\partial t \partial \xi_u} & \frac{\partial^2 S}{\partial t \partial t} \end{pmatrix} . \tag{3.88}$$

The first factor can be expressed in terms of the monodromy matrix in the following way. If we write the monodromy matrix in $(d-1) \times (d-1)$ blocks,

$$\mathbb{M}_\gamma =: \begin{pmatrix} A & B \\ C & D \end{pmatrix} , \tag{3.89}$$

then we can deduce from the definition of $\mathbb{M}_\gamma$,

$$\begin{pmatrix} \frac{\partial S}{\partial x_v} \\ x_v \end{pmatrix} = \mathbb{M}_\gamma \begin{pmatrix} \xi_v \\ \frac{\partial S}{\partial \xi_v} \end{pmatrix} , \tag{3.90}$$

the identities

$$\begin{aligned} \frac{\partial S}{\partial \xi_v}(\boldsymbol{x}, \boldsymbol{\xi}, t) &= D^{-1} x_v - D^{-1} C \xi_v \\ \frac{\partial S}{\partial x_v}(\boldsymbol{x}, \boldsymbol{\xi}, t) &= (A - BD^{-1}C)\xi_v + BD^{-1} x_v . \end{aligned} \tag{3.91}$$

By taking further derivatives the first term of (3.88) reads

$$\begin{aligned} &\det \begin{pmatrix} \frac{\partial^2 S}{\partial x_v \partial x_v} & \frac{\partial^2 S}{\partial x_v \partial \xi_v} - \mathbb{1}_{d-1} \\ \frac{\partial^2 S}{\partial \xi_v \partial x_v} - \mathbb{1}_{d-1} & \frac{\partial^2 S}{\partial \xi_v \partial \xi_v} \end{pmatrix} \\ &= \det \begin{pmatrix} BD^{-1} & A - BD^{-1}C - \mathbb{1}_{d-1} \\ D^{-1} - \mathbb{1}_{d-1} & -D^{-1}C \end{pmatrix} \\ &= \det B \ \det \begin{pmatrix} D^{-1} & B^{-1}A - D^{-1}C - B^{-1} \\ D^{-1} - \mathbb{1}_{d-1} & -D^{-1}C \end{pmatrix} \end{aligned} \tag{3.92}$$

Subtracting the second line from the first line we obtain

$$\begin{aligned} &\det \begin{pmatrix} \frac{\partial^2 S}{\partial x_v \partial x_v} & \frac{\partial^2 S}{\partial x_v \partial \xi_v} - \mathbb{1}_{d-1} \\ \frac{\partial^2 S}{\partial \xi_v \partial x_v} - \mathbb{1}_{d-1} & \frac{\partial^2 S}{\partial \xi_v \partial \xi_v} \end{pmatrix} \\ &= \det B \ \det \begin{pmatrix} \mathbb{1}_{d-1} & B^{-1}A - B^{-1} \\ D^{-1} - \mathbb{1}_{d-1} & -D^{-1}C \end{pmatrix} \\ &= \det B \ \det D^{-1} \ \det \begin{pmatrix} \mathbb{1}_{d-1} & B^{-1}A - B^{-1} \\ \mathbb{1}_{d-1} - D & -C \end{pmatrix} \\ &= \det D^{-1} \ \det \begin{pmatrix} B & A - \mathbb{1}_{d-1} \\ \mathbb{1}_{d-1} - D & -C \end{pmatrix} . \end{aligned} \tag{3.93}$$

Inverting the sign in the second column and exchanging the first and second column we finally get

$$\det\begin{pmatrix} \dfrac{\partial^2 S}{\partial x_v \partial x_v} & \dfrac{\partial^2 S}{\partial x_v \partial \xi_v} - \mathbb{1}_{d-1} \\ \dfrac{\partial^2 S}{\partial \xi_v \partial x_v} - \mathbb{1}_{d-1} & \dfrac{\partial^2 S}{\partial \xi_v \partial \xi_v} \end{pmatrix} = \det D^{-1}\ \det(\mathbb{M}_\gamma - \mathbb{1}_{2d-2})$$
$$= \det\left(\frac{\partial^2 S}{\partial x_v \partial \xi_v}\right)\ \det(\mathbb{M}_\gamma - \mathbb{1}_{2d-2})\ . \tag{3.94}$$

We now turn to the evaluation of the second factor of (3.88). From (3.81) we have the relation

$$\begin{pmatrix} \frac{\partial S}{\partial x_u} \\ x_u \end{pmatrix} = \mathbb{1}_2 \begin{pmatrix} \xi_u \\ \frac{\partial S}{\partial \xi_u} \end{pmatrix}\ , \tag{3.95}$$

for $t = T_\gamma$, which implies

$$\frac{\partial^2 S}{\partial \xi_u^2} = 0 = \frac{\partial^2 S}{\partial x_u^2} \quad \text{and} \quad \frac{\partial^2 S}{\partial x_u \partial \xi_u} = 1\ . \tag{3.96}$$

Before we can use these identities (recall that they only hold for $t = T_\gamma$) we first have to rewrite the time derivatives using the Hamilton–Jacobi equation,

$$\frac{\partial^2 S}{\partial \xi_u \partial t} = -\frac{\partial}{\partial \xi_u} H\left(\frac{\partial S}{\partial x}, \boldsymbol{x}\right) = \frac{\partial H}{\partial p_u}\frac{\partial^2 S}{\partial x_u \partial \xi_u} - -\dot{X}_u\ . \tag{3.97}$$

Note that we have again used the separation (3.82). The second factor of (3.81) now reads

$$\det\begin{pmatrix} \dfrac{\partial^2 S}{\partial \xi_u \partial \xi_u} & \dfrac{\partial^2 S}{\partial \xi_u \partial t} \\ \dfrac{\partial^2 S}{\partial t \partial \xi_u} & \dfrac{\partial^2 S}{\partial t \partial t} \end{pmatrix} = \det\begin{pmatrix} 0 & -\dot{X}_u \\ -\dot{X}_u & \dfrac{\partial^2 S}{\partial t \partial t} \end{pmatrix} = -\dot{X}_u^2 \tag{3.98}$$

Thus, for the relevant determinant we finally get

$$\det D^2\phi = -\dot{X}_u^2\ \det\left(\frac{\partial^2 S}{\partial x_v \partial \xi_v}\right)\ \det(\mathbb{M}_\gamma - \mathbb{1}_{2d-2})$$
$$= -\dot{X}_u^2\ \det\left(\frac{\partial^2 S}{\partial x \partial \xi}\right)\ \det(\mathbb{M}_\gamma - \mathbb{1}_{2d-2})\ , \tag{3.99}$$

where in the last step we have used (3.96)

When we now perform the stationary phase integrals in (3.66) the second factor cancels with the respective determinant factor in a_0, cf. (3.41), and for the contribution to the trace formula we find

$$\mathcal{A}_\gamma = \frac{\mathrm{e}^{-\mathrm{i}(\pi/2)\mu_\gamma}}{|\det(\mathbb{M}_\gamma - \mathbb{1}_{2d-2})|^{1/2}} \oint_\gamma \frac{1}{|\dot{X}_u|} \operatorname{tr} D(\boldsymbol{\xi}, \boldsymbol{y}, T_\gamma)\, \mathrm{d}X_u \left[1 + \mathcal{O}(\hbar)\right] . \quad (3.100)$$

The action S_γ can be expressed as

$$S_\gamma = \oint_\gamma \boldsymbol{P}\, \mathrm{d}\boldsymbol{X} \quad (3.101)$$

and the factor $\exp(-\mathrm{i}(\pi/2)\mu_\gamma)$ keeps track of all phase factors from the method of stationary phase and those inherent in a_0. It can be shown that μ_γ is the Maslov index of the orbit counting the maximal number of conjugate points along γ, see [6–8, 21]. More information on the geometrical meaning of the Maslov index and how to calculate it can, e.g., be found in [39, 40].

In Sect. 3.4 we have seen that the spin contribution $\operatorname{tr} D(\boldsymbol{\xi}, \boldsymbol{y}, T_\gamma)$, $(\boldsymbol{\xi}, \boldsymbol{y}) \in \gamma$, appearing in the remaining integral is independent of the initial point $(\boldsymbol{\xi}, \boldsymbol{y})$ on the orbit, see (3.52), and thus can be pulled out of the integral. Therefore, we only have to calculate

$$\oint_\gamma \frac{\mathrm{d}X_u}{|\dot{X}_u|} , \quad (3.102)$$

which, with $\mathrm{d}X_u/\dot{X}_u = \mathrm{d}t$ yields the primitive period $T_\gamma^\#$ of the orbit. Note that, since the integral is a remnant of the trace integral, it extends over all points in configuration space which lie on the periodic orbit γ. In particular, if γ is a k-fold repetition of a primitive orbit $\gamma^\#$ all points on the orbit are only counted once instead of k times.

Finally, we can now write down the contribution to (3.67) of an isolated and unstable period orbit γ,

$$\mathcal{A}_\gamma = \frac{T_\gamma^\#\, \chi_s(\alpha_\gamma)}{|\det(\mathbb{M}_\gamma - \mathbb{1}_{2d-2})|^{1/2}}\, \mathrm{e}^{-\mathrm{i}(\pi/2)\mu_\gamma} \left[1 + \mathcal{O}(\hbar)\right] , \quad (3.103)$$

where χ_s denotes the character of the $(2s+1)$-dimensional unitary irreducible representation of SU(2) and α_γ is the angle by which a classical spin vector has been rotated after precession along the periodic orbit γ, cf. Sect. 3.4 and Appendix C.

The trace formula for a strongly chaotic system, for which the flow ϕ_H^t has only isolated and unstable periodic orbits therefore reads

$$\begin{aligned} \sum_n \varrho\left(\frac{E_n - E}{\hbar}\right) &= \frac{\hat{\varrho}(0)}{2\pi} \frac{(2s+1)\,|\Omega_E|}{(2\pi\hbar)^{d-1}} \left[1 + \mathcal{O}(\hbar)\right] \\ &+ \sum_\gamma \frac{\hat{\varrho}(T_\gamma)}{2\pi} \frac{T_\gamma^\#\, \chi_s(\alpha_\gamma)}{|\det(\mathbb{M}_\gamma - \mathbb{1}_{2d-2})|^{1/2}} \exp\left(\frac{\mathrm{i}}{\hbar} S_\gamma - \mathrm{i}\frac{\pi}{2}\mu_\gamma\right) \left[1 + \mathcal{O}(\hbar)\right] , \end{aligned} \quad (3.104)$$

where the sum on the right hand side extends over all periodic orbits with energy E. We can also express this result in terms of the respective properties of primitive periodic orbits $\gamma^{\#}$ and their k-fold repetitions, formally including negative ones, by

$$\sum_n \varrho\left(\frac{E_n - E}{\hbar}\right) = \frac{\hat{\varrho}(0)}{2\pi}\frac{(2s+1)\,|\Omega_E|}{(2\pi\hbar)^{d-1}}\,[1 + \mathcal{O}(\hbar)]$$
$$+ \sum_{\gamma^{\#}} \sum_{k\in\mathbb{Z}\setminus\{0\}} \frac{\hat{\varrho}(kT_{\gamma^{\#}})}{2\pi}\frac{T_{\gamma^{\#}}\,\chi_s(k\alpha_{\gamma^{\#}})}{|\det(\mathbb{M}^k_{\gamma^{\#}} - \mathbb{1}_{2d-2})|^{1/2}} \qquad (3.105)$$
$$\times \exp\left(\frac{\mathrm{i}}{\hbar}kS_{\gamma^{\#}} - \mathrm{i}\frac{\pi}{2}k\mu_{\gamma^{\#}}\right)\,[1 + \mathcal{O}(\hbar)]\,.$$

For spin $s = 0$ this is (a regularised version of) the Gutzwiller trace formula as originally stated in [9]. For spin $s = \frac{1}{2}$ (either described by the Pauli or the Dirac equation) the respective result was derived in [1–3].

3.5.3 Integrable Systems

If the classical translational dynamics, i.e. the flow ϕ^t_H, is integrable we can locally introduce action and angle variables $(\boldsymbol{I}, \boldsymbol{\vartheta})$ on phase space by a canonical transformation, see e.g. [35, 36] and Sect. 4.4.1. The new classical Hamiltonian, which we will denote by $\overline{H}$, is a function of the action variables only. The equations of motion become particularly simple, since all $\boldsymbol{I}$ are conserved and

$$\dot{\boldsymbol{\vartheta}} = \nabla_{\boldsymbol{I}}\overline{H}(\boldsymbol{I}) =: \boldsymbol{\omega}(\boldsymbol{I}) = const\,. \qquad (3.106)$$

For a fixed set of action variables $\boldsymbol{I}$ the dynamics takes place on a d-torus $\mathbb{T}^d$, i.e. the angles are defined modulo 2π. The generating function $S(\boldsymbol{x}, \boldsymbol{\xi}, t)$ in action and angle variables, i.e. $S(\boldsymbol{\vartheta}, \boldsymbol{I}, t)$, has to fulfill

$$\nabla_{\boldsymbol{\vartheta}} S = \boldsymbol{I}\,, \quad \nabla_{\boldsymbol{I}} S = \boldsymbol{\vartheta}_0 \quad \text{and} \quad \frac{\partial S}{\partial t} = -E = -\overline{H}(\boldsymbol{I})\,, \qquad (3.107)$$

for a motion from $\boldsymbol{\vartheta}_0$ to $\boldsymbol{\vartheta}$ in time t. One easily verifies that

$$S(\boldsymbol{\vartheta}, \boldsymbol{I}, t) := \boldsymbol{I}\boldsymbol{\vartheta} - \overline{H}(\boldsymbol{I})t \qquad (3.108)$$

is a suitable choice. Then the phase in the semiclassical ansatz (3.29) is given by

$$\boldsymbol{I}(\boldsymbol{\vartheta} - \boldsymbol{\vartheta}_0) - \overline{H}(\boldsymbol{I})t\,, \qquad (3.109)$$

and for the leading order amplitude one finds $a_0(\boldsymbol{I}, \boldsymbol{\vartheta}, t) = D(\boldsymbol{I}, \boldsymbol{\vartheta}, t)$, since

$$\det\frac{\partial^2 S}{\partial\vartheta\partial I} = \det\mathbb{1}_d = 1\,. \qquad (3.110)$$

Hence, when transforming (3.66) to action and angle variables we have to replace the total phase $S(\boldsymbol{x}, \boldsymbol{\xi}, t) - \boldsymbol{y}\boldsymbol{\xi} + Et$ by $\boldsymbol{I}(\boldsymbol{\vartheta} - \boldsymbol{\vartheta}_0) - \overline{H}(\boldsymbol{I})t + Et$.

The equations of motion (3.106) allow periodic solutions only on tori whose frequencies $\boldsymbol{\omega}(\boldsymbol{I})$ are rationally dependent, more precisely,

$$\frac{\omega_j}{\omega_k} \in \mathbb{Q} \quad \forall\, j,k = 1,\ldots,d\,. \tag{3.111}$$

Consequently for some time t we have

$$\boldsymbol{\omega} t = 2\pi \boldsymbol{m}\,, \qquad \boldsymbol{m} \in \mathbb{Z}^d \setminus \{0\}\,. \tag{3.112}$$

We will label all quantities on the corresponding torus by $\boldsymbol{m}$, the vector of winding numbers. The time which solves (3.112) we denote by

$$T_{\boldsymbol{m}} := \frac{2\pi}{\omega_j} m_j \quad \forall\, j = 1,\ldots,d\,, \tag{3.113}$$

and for the actions which characterise the respective torus we write $\boldsymbol{I}_{\boldsymbol{m}}$. We will also label the contribution to the trace formula (3.67) deriving from a rational torus with winding numbers $\boldsymbol{m}$ by $\mathcal{A}_{\boldsymbol{m}}$. Such a contribution can be calculated directly, as we will demonstrate below.

Note, that by the above procedure we have implicitly assumed that the classical dynamics is not degenerate, i.e. that for a given $\boldsymbol{m}$ there is at most one set of actions $\boldsymbol{I}$, for which (3.112) can be fulfilled. Otherwise our labeling would not be unique. However, even if there is more than one solution $\boldsymbol{I}$ to (3.112) for fixed $\boldsymbol{m}$, the calculation for a single torus remains the same, as long as there is no continuous family of actions $\boldsymbol{I}$ for which (3.112) can be fulfilled with a fixed $\boldsymbol{m}$. In the latter case the classical dynamics is degenerate and the sets $\mathcal{M}_s$ of stationary points of the total phase in (3.66) have dimension greater than d. We will not discuss this situation here but remark that the respective contributions to the trace formula can still be calculated along the same lines as presented here and in the previous sections. A general method for deriving the contribution of families of periodic orbits, for which the family is generated by the action of a symmetry group on a member of the family was given in [38].

We now want to calculate the contribution of a single rational torus with winding numbers $\boldsymbol{m}$. After we have transformed to action and angle variables in (3.66), we can perform the integrations over t and $\boldsymbol{I}$ by the method of stationary phase, keeping only the $\boldsymbol{\vartheta}$-integrations to be performed exactly. To this end we need the second derivative of the total phase

$$\phi := 2\pi \boldsymbol{I}\boldsymbol{m} + (E - \overline{H}(\boldsymbol{I}))t\,, \tag{3.114}$$

where we have used, that in order to take the trace for the translational degrees of freedom we have to set $\boldsymbol{\vartheta} = \boldsymbol{\vartheta}_0 + 2\pi\boldsymbol{m}$ instead of merely choosing $\boldsymbol{\vartheta} = \boldsymbol{\vartheta}_0$. The relevant matrix now reads

$$\mathbb{D}_{\boldsymbol{m}} := \left.\begin{pmatrix} \dfrac{\partial^2\phi}{\partial t^2} & \dfrac{\partial^2\phi}{\partial t\partial I} \\[2ex] \dfrac{\partial^2\phi}{\partial t\partial I} & \dfrac{\partial^2\phi}{\partial I\partial I} \end{pmatrix}\right|_{\boldsymbol{I}=\boldsymbol{I}_{\boldsymbol{m}},t=T_{\boldsymbol{m}}} = \begin{pmatrix} 0 & \boldsymbol{\omega}^T(\boldsymbol{I}_{\boldsymbol{m}}) \\[2ex] \boldsymbol{\omega}(\boldsymbol{I}_{\boldsymbol{m}}) & T_{\boldsymbol{m}}\dfrac{\partial^2\overline{H}}{\partial I\partial I}(\boldsymbol{I}_{\boldsymbol{m}}) \end{pmatrix}. \tag{3.115}$$

If one correctly keeps track of the additional phases occurring in the calculation, which we will not show here, the resulting phase factor is given by

$$\exp\left(-\mathrm{i}\frac{\pi}{2}\boldsymbol{m}\boldsymbol{\mu}-\mathrm{i}\frac{\pi}{4}(d-1)\mathrm{sign}T_{\boldsymbol{m}}\right) . \tag{3.116}$$

Here μ_j denotes the Maslov index of the jth basis cycle $\mathcal{C}_j$ on $\mathbb{T}^d$, which can, for instance, be realised by a loop along which all ϑ_k, $k \neq j$ are held constant, see also Sect. 4.4.1. Thus we immediately find

$$\mathcal{A}_{\boldsymbol{m}} = \frac{\exp\left(-\mathrm{i}\frac{\pi}{2}\boldsymbol{m}\boldsymbol{\mu}-\mathrm{i}\frac{\pi}{4}(d-1)\mathrm{sign}T_{\boldsymbol{m}}\right)}{(2\pi\hbar)^{(d-1)/2}\left|\det \mathbb{D}_{\boldsymbol{m}}\right|^{1/2}} \underbrace{\int_{\mathbb{T}^d} \mathrm{tr}\, D(\boldsymbol{I}_{\boldsymbol{m}},\boldsymbol{\vartheta},T_{\boldsymbol{m}})\,\mathrm{d}^d\vartheta}_{=:(2\pi)^d\,\overline{\chi_s}^{\boldsymbol{m}}} \left[1+\mathcal{O}(\hbar)\right] . \tag{3.117}$$

Note, that the spin-weight which appears under the remaining integral is not constant on the torus, but depends on $\boldsymbol{\vartheta}$. From general considerations we know, see Sect. 3.4, that for any two points which lie on the same periodic orbit the factor $\mathrm{tr}\, D$ evaluated at the period of the orbit is identical. Thus, we could introduce new coordinates on $\mathbb{T}^d$ such that only ϑ_1 varies along the flow lines of ϕ^t_H and therefore the ϑ_1-integration yields a trivial factor of 2π. However the remaining $(d-1)$ integrals are non-trivial. Only if we impose additional conditions on the spin dynamics, we will be able to perform these integrations. Since this requires a deeper insight into the geometrical structures underlying the combined dynamics of classical translational motion and classical spin precession we will postpone this question until Sect. 4.5.

Together with the result from Sect. 3.5.1 we can now write down the trace formula for spin-s systems with integrable translational dynamics,

$$\begin{aligned}\sum_n \varrho\left(\frac{E_n-E}{\hbar}\right) &= \frac{\hat{\varrho}(0)}{2\pi}\frac{(2s+1)\left|\Omega_E\right|}{(2\pi\hbar)^{d-1}}\left[1+\mathcal{O}(\hbar)\right] \\ &+ \sum_{\boldsymbol{m}\in\mathbb{Z}^d\setminus\{0\}} \frac{\hat{\varrho}(T_{\boldsymbol{m}})}{2\pi}\frac{(2\pi)^d\exp\left(-\mathrm{i}\frac{\pi}{2}\boldsymbol{m}\boldsymbol{\mu}-\mathrm{i}\frac{\pi}{4}(d-1)\mathrm{sign}T_{\boldsymbol{m}}\right)}{(2\pi\hbar)^{(d-1)/2}\left|\det\mathbb{D}_{\boldsymbol{m}}\right|^{1/2}} \\ &\qquad \times \mathrm{e}^{(\mathrm{i}/\hbar)2\pi\boldsymbol{m}\boldsymbol{I}}\,\overline{\chi_s}^{\boldsymbol{m}}\left[1+\mathcal{O}(\hbar)\right] .\end{aligned} \tag{3.118}$$

Notice, that as in the previously discussed case of isolated orbits the rapidly oscillating term involves the action $S = 2\pi\boldsymbol{m}\boldsymbol{I}$ of a periodic orbit on $\mathbb{T}^d$, since the action variable

$$I_j = \frac{1}{2\pi}\oint_{\mathcal{C}_j} \boldsymbol{p}\,\mathrm{d}\boldsymbol{x} \tag{3.119}$$

is the action along the jth basis cycle, which in (3.118) is multiplied by the respective winding number m_j.

Often it is convenient to write the result not in terms of complex exponentials but by using trigonometrical functions instead. To this end notice that for $\boldsymbol{m} \mapsto -\boldsymbol{m}$ we have

$$T_{\boldsymbol{m}} \mapsto -T_{\boldsymbol{m}}\,, \quad |\det D_{\boldsymbol{m}}| \mapsto |\det D_{\boldsymbol{m}}| \quad \text{and} \quad \overline{\chi_s}^{\boldsymbol{m}} \mapsto \overline{\chi_s}^{\boldsymbol{m}}\,. \tag{3.120}$$

The last property is true, since all angles $\alpha(\boldsymbol{\vartheta}, T_{\boldsymbol{m}})$, by which a spin vector is rotated when transported along a trajectory starting at $\boldsymbol{\vartheta}$ up to time $T_{\boldsymbol{m}}$, are inverted, which, however, does not change the character $\chi_s(\alpha(\boldsymbol{\vartheta}, T_{\boldsymbol{m}}))$. Thus, for symmetric test functions $\hat{\varrho}$ the trace formula (3.118) can be written as

$$\begin{aligned} \sum_n \varrho\left(\frac{E_n - E}{\hbar}\right) &= \frac{\hat{\varrho}(0)}{2\pi}\,\frac{(2s+1)\,|\Omega_E|}{(2\pi\hbar)^{d-1}}\,[1 + \mathcal{O}(\hbar)] \\ &+ \sum_{\boldsymbol{m}\in\mathbb{Z}^d\setminus\{0\}} \frac{\hat{\varrho}(T_{\boldsymbol{m}})}{2\pi}\,\frac{(2\pi)^d\,\overline{\chi_s}^{\boldsymbol{m}}}{(2\pi\hbar)^{(d-1)/2}\,|\det \mathbb{D}_{\boldsymbol{m}}|^{1/2}} \times \\ &\times \cos\left(\frac{2\pi \boldsymbol{m}\boldsymbol{I}}{\hbar} - \frac{\pi}{2}\boldsymbol{m}\boldsymbol{\mu} - \frac{\pi}{4}(d-1)\right)[1 + \mathcal{O}(\hbar)]\,. \end{aligned} \tag{3.121}$$

3.6 Examples

We illustrate the general results obtained in the preceeding sections for some example systems. First we briefly revisit our introductory example from Chap. 2, before we discuss spin–orbit coupling in billiards and in the hydrogen atom.

3.6.1 Reprise: Oscillators

After having seen the derivation of trace formulae for systems with spin in general we briefly return to our toy example of Chap. 2. There we investigated a harmonic oscillator system with spin 1/2, cf. (2.17). Being a one dimensional system the (exact) trace formula (which we derived by means of Poisson summation) can be seen as a special case of both the general result for systems with only isolated periodic orbits (3.104) as well as of the expression (3.118) for integrable systems. In Sect. 2.1 we have already given a preliminary comparison with the general results. Therefore, we only list the following quantities in order to complete the discussion.

The action of the (primitive) periodic orbit of a harmonic oscillator is given by

$$S = \frac{2\pi E}{\omega}\,, \tag{3.122}$$

its period and Maslov index read

$$T = \frac{2\pi}{\omega} \quad \text{and} \quad \mu = 2\,, \tag{3.123}$$

respectively (one encounters two turning points along the orbit). For a system with one translational degree of freedom there is no monodromy matrix (in the language of Sect. 3.5.2 there are no coordinates $\boldsymbol{X}_v$), i.e. when inserting the previously listed quantities into (3.104) the term $|\det \mathbb{M}_\gamma^k - \mathbb{1}|$ has to be omitted. The reader easily verifies that, together with the spin contribution discussed in Sect. 2.1 and the formal choice $\varrho(\omega) = \delta(\hbar\omega)$ for the test function, these substitutions turn (3.104) into (2.21).

On the other hand the Hamiltonian of a harmonic oscillator expressed in action and angle variables reads $\overline{H}(I) = \omega I$, see e.g. [35], i.e.

$$|\det \mathbb{D}|^{1/2} = \omega \,, \tag{3.124}$$

cf. (3.115). Again one easily confirms that the respective substitutions turn (3.118) into (2.21).

3.6.2 Spin–Orbit Coupling in 2 Dimensions – $\boldsymbol{\sigma p}$-Billiards

We will introduce a class of systems which will be used in order to illustrate various aspects of semiclassics and spectral statistics for systems with spin.

In the study of quantum chaos for spinless particles so-called billiard systems are useful and popular examples. In this context a classical two-dimensional Euclidean billiard is defined by a compact domain $D \subset \mathbb{R}^2$ with boundary ∂D. Inside the billiard one considers the motion of a free particle with elastic reflections at the boundary. The corresponding quantum system is given the eigenvalue problem of the negative Laplacian $-\Delta$ on $L^2(D)$ with suitable boundary conditions on ∂D, often chosen to be Dirichlet.

If we want to study the influence of spin in such a system, we have to modify the Schrödinger operator

$$\hat{H} = -\frac{\hbar^2}{2m}\Delta \,, \tag{3.125}$$

where m denotes the mass of the particle, by adding a term which couples the translational and spin degrees of freedom. We will choose to investigate Pauli operators of the form

$$\hat{H}_{\mathrm{P}} = -\frac{\hbar^2}{2m}\Delta + \frac{\hbar}{2}\kappa\, \mathrm{d}\pi_s(\boldsymbol{\sigma})\hat{\boldsymbol{p}} \,, \tag{3.126}$$

where $\hat{\boldsymbol{p}} = (\hbar/\mathrm{i})\nabla$ is the momentum operator and κ is some fixed coupling constant. Since the additional term couples the spin operator $\hat{\boldsymbol{s}} = (\hbar/2)\mathrm{d}\pi_s(\boldsymbol{\sigma})$ to the translational motion we call these systems billiards with spin–orbit coupling or "$\boldsymbol{\sigma p}$-billiards".

The coupling term can indeed be derived from the well-known expression for spin–orbit coupling in three dimensions as follows. In a non-relativistic context spin–orbit coupling is described by a Pauli equation whose spin-dependent term reads

$$\frac{e}{2m^2c^2}(\hat{\boldsymbol{p}} \times \boldsymbol{E})\,\hat{\boldsymbol{s}} \tag{3.127}$$

where $\boldsymbol{E}$ denotes an external electric field. This term is the non-relativistic limit of the respective contribution which can be derived from the Dirac equation, see e.g. [28, 29, 41]. If we want to confine particles to, say, the xy-plane, we have to apply a steep potential ϕ in the z-direction. Concerning the motion in z-direction the system will almost always be in the ground state for small energies, but arbitrary excitations in x- and y-direction are allowed. Therefore, such a system is called quasi two-dimensional. This method is used experimentally to produce quasi two-dimensional semiconductor quantum structures, so-called quantum wells. Hence, we have to consider (3.127) with $|E_z| \gg |E_x|, |E_y|$, in which case we make the approximation

$$\boldsymbol{E} \times \hat{\boldsymbol{p}} \approx \begin{pmatrix} E_z \hat{p}_y \\ -E_z \hat{p}_x \\ 0 \end{pmatrix} . \tag{3.128}$$

Absorbing E_z together with the pre-factor of (3.127) in a coupling constant κ, the spin–orbit term in two dimensions reads

$$\kappa \left(\hat{p}_y \frac{\hbar}{2} \mathrm{d}\pi_s(\sigma_x) - \hat{p}_x \frac{\hbar}{2} \mathrm{d}\pi_s(\sigma_y) \right) . \tag{3.129}$$

Now we can always apply a unitary transformation in spin space, namely a rotation $\exp(-\mathrm{i}\pi\sigma_z/4)$ about the σ_z-axis by $\pi/2$, which maps

$$\sigma_x \mapsto \mathrm{e}^{-\mathrm{i}\pi\sigma_z/4}\, \sigma_x\, \mathrm{e}^{\mathrm{i}\pi\sigma_z/4} = \sigma_y \quad \text{and} \quad \sigma_y \mapsto -\sigma_x . \tag{3.130}$$

Finally, we have found the spin–orbit coupling term,

$$\kappa \hat{\boldsymbol{p}} \frac{\hbar}{2} \mathrm{d}\pi_s(\boldsymbol{\sigma}) , \tag{3.131}$$

as stated in (3.126). We remark that in semiconductor physics a term of this form, known as the Rashba term, see e.g. [42], is indeed used for modeling spin–orbit coupling. The methods developed in this thesis can be successfully applied in this context [43]. However, one should note that the derivation of (3.131) is much more involved and does not follow the simple lines shown here. We refrain from going into details on this problem, but refer the interested reader to the literature, see e.g. [44, 45]. For our purposes, however, the above discussion should be a sufficient motivation for the study of $\boldsymbol{\sigma p}$-billiards.

3.6.3 The $\boldsymbol{\sigma p}$-Torus

As a specific example let us consider the Hamiltonian (3.126) for a rectangle with lengths a_1 and a_2,

$$D = \left\{ \boldsymbol{x} \in \mathbb{R}^2 \,|\, 0 \le x_j \le a_j \right\} , \tag{3.132}$$

and periodic boundary conditions, i.e we put the system on a torus. For simplicity we chose $2m = 1$. The results of the preceeding sections require that we search for periodic orbits of the classical Hamiltonian

$$H(\boldsymbol{p}, \boldsymbol{x}) = \boldsymbol{p}^2 \tag{3.133}$$

and solve the following equation of spin precession along these orbits,

$$\dot{\boldsymbol{s}} = \kappa \, \boldsymbol{p} \times \boldsymbol{s} \, . \tag{3.134}$$

To this end notice that $H(\boldsymbol{p}, \boldsymbol{x})$ is already given in action and angle variables, since H does not depend on $\boldsymbol{x}$ and the set of phase space with a fixed value of $\boldsymbol{p}$ is the torus $\mathbb{R}^2/(a_1\mathbb{Z} \times a_2\mathbb{Z})$. If we rescale the coordinates such that the lengths of the rectangle become 2π, i.e. we define

$$\vartheta_j := \frac{2\pi}{a_j} x_j \, , \tag{3.135}$$

we also have to rescale the momenta according to

$$I_j := \frac{a_j}{2\pi} p_j \tag{3.136}$$

in order to make the transformation canonical, i.e. area preserving in phase space. Thus, we obtain the new Hamiltonian

$$\overline{H}(\boldsymbol{I}) = 4\pi^2 \left(\frac{I_1^2}{a_1^2} + \frac{I_2^2}{a_2^2} \right) \tag{3.137}$$

and the frequencies

$$\boldsymbol{\omega} = 8\pi^2 \begin{pmatrix} I_1/a_1^2 \\ I_2/a_2^2 \end{pmatrix} \, . \tag{3.138}$$

We find periodic orbits with period $T_{\boldsymbol{m}}$ for actions $\boldsymbol{I}$ fulfilling

$$T_{\boldsymbol{m}} = \frac{2\pi}{\omega_j} m_j = \frac{a_j^2 m_j}{4\pi I_j} \quad \Rightarrow \quad I_j = \frac{a_j^2 m_j}{4\pi T_{\boldsymbol{m}}} \, . \tag{3.139}$$

Note, that for each pair $(m_1, m_2) \neq (0, 0)$ we can find two solutions $T_{\boldsymbol{m}}$, one of them positive, the other negative. In the following we will denote by $T_{\boldsymbol{m}}$ the positive solution, but when writing down the trace formula below, we will have to include an additional factor of 2. The energy of these periodic orbits is given by

$$E = H\left(\frac{a_1^2 m_1}{4\pi T_{\boldsymbol{m}}}, \frac{a_2^2 m_2}{4\pi T_{\boldsymbol{m}}} \right) = \frac{a_1^2 m_1^2 + a_2^2 m_2^2}{4 T_{\boldsymbol{m}}^2} \, , \tag{3.140}$$

allowing us to write the period $T_{\boldsymbol{m}}$ as a function of the integers $\boldsymbol{m}$ and the energy E,

$$T_{\boldsymbol{m}} = \frac{1}{2\sqrt{E}} \sqrt{a_1^2 m_1^2 + a_2^2 m_2^2} \, . \tag{3.141}$$

Similarly we find their action by

$$S_{\boldsymbol{m}} = 2\pi \boldsymbol{m}\boldsymbol{I} = \sqrt{E}\sqrt{a_1^2 m_1^2 + a_2^2 m_2^2}\,. \tag{3.142}$$

Notice that the last equation can also be written in the form

$$S_{\boldsymbol{m}} = |\boldsymbol{p}| l_{\boldsymbol{m}}\,, \tag{3.143}$$

where $|\boldsymbol{p}| := \sqrt{E}$ is the conserved absolute value of momentum, and $l_{\boldsymbol{m}} := \sqrt{a_1^2 m_1^2 + a_2^2 m_2^2}$ denotes the length of the orbit. Such a relation is always true for billiard systems. With

$$\frac{\partial \omega_j}{\partial I_k} = \frac{8\pi^2}{a_j^2}\delta_{jk} \qquad \omega_j = \frac{8\pi^2}{a_j^2} I_j = \frac{2\pi m_j}{T_{\boldsymbol{m}}} \tag{3.144}$$

we can also calculate the determinant of the matrix (3.115),

$$\begin{aligned} \det \mathbb{D}_{\boldsymbol{m}} &= \det \begin{pmatrix} 0 & \frac{2\pi m_1}{T_{\boldsymbol{m}}} & \frac{2\pi m_2}{T_{\boldsymbol{m}}} \\ \frac{2\pi m_1}{T_{\boldsymbol{m}}} & \frac{8\pi^2}{a_1^2} T_{\boldsymbol{m}} & 0 \\ \frac{2\pi m_2}{T_{\boldsymbol{m}}} & 0 & \frac{8\pi^2}{a_2^2} T_{\boldsymbol{m}} \end{pmatrix} = -\frac{32\pi^4}{T_{\boldsymbol{m}}}\left(\frac{m_1^2}{a_2^2} + \frac{m_2^2}{a_1^2}\right) \\ &= -\frac{64\pi^4}{a_1^2 a_2^2}\sqrt{E}\, l_{\boldsymbol{m}}\,. \end{aligned} \tag{3.145}$$

In order to complete the information required for the trace formula we still need the volume of the energy shell,

$$|\Omega_E| = \int_D \int_{\mathbb{R}^2} \delta(\boldsymbol{p}^2 - E)\, \mathrm{d}^2 p\, \mathrm{d}^2 x = \pi |D|, \tag{3.146}$$

where $|D| = a_1 a_2$ is the area of the billiard, and finally, the spin contribution. For the latter we have to solve the equation of spin precession (3.134) along each periodic orbit of a rational torus until time $T_{\boldsymbol{m}}$. However, since the momentum $\boldsymbol{p}$ is conserved along the orbit, any initial spin vector $\boldsymbol{s}$ is rotated by the angle

$$\alpha_{\boldsymbol{m}} = \kappa |\boldsymbol{p}| T_{\boldsymbol{m}} = \frac{\kappa}{2} l_{\boldsymbol{m}} \tag{3.147}$$

about an axis parallel to $\boldsymbol{p}$. This result is independent of the initial point $\boldsymbol{x}$ on the torus but only depends on the length $l_{\boldsymbol{m}}$ of the periodic orbit. Thus the spin contribution reads

$$\overline{\chi_s}^{\boldsymbol{m}} = \chi_s(\alpha_{\boldsymbol{m}}) = \sum_{m=-s}^{s} \exp\left(\mathrm{i} m \frac{\kappa}{2} l_{\boldsymbol{m}}\right)\,, \tag{3.148}$$

where we have used formula (3.63) for the character χ_s of the $(2s+1)$ dimensional unitary irreducible representation of SU(2).

Finally, inserting everything into (3.121), the semiclassical trace formula for the $\boldsymbol{\sigma p}$-torus reads

$$\sum_n \varrho\left(\frac{E_n - E}{\hbar}\right) = \frac{\hat{\varrho}(0)}{2\pi}\frac{(2s+1)|D|}{2\hbar}\left[1 + \mathcal{O}(\hbar)\right]$$
$$+ \sum_{\boldsymbol{m}\in\mathbb{Z}^2\setminus\{0\}} \frac{\hat{\varrho}(T_{\boldsymbol{m}})}{2\pi}\frac{|D|\chi_s(\alpha_{\boldsymbol{m}})}{\sqrt{2\pi\hbar l_{\boldsymbol{m}}\sqrt{E}}}\cos\left(\frac{\sqrt{E}}{\hbar}l_{\boldsymbol{m}} - \frac{\pi}{4}\right)\left[1 + \mathcal{O}(\hbar)\right], \tag{3.149}$$

with $T_{\boldsymbol{m}} = l_{\boldsymbol{m}}/(2\sqrt{E})$, $\alpha_{\boldsymbol{m}} = \kappa l_{\boldsymbol{m}}/2$ and $l_{\boldsymbol{m}} = \sqrt{a_1^2 m_1^2 + a_2^2 m_2^2}$. We can also obtain a semiclassical trace formula for the spectral density $d(E) = \sum_n \delta(E - E_n)$, formally by choosing $\varrho(\omega) = \delta(\hbar\omega)$, which reads

$$d(E) = (2s+1)\frac{\pi|D|}{(2\pi\hbar)^2}\left[1 + \mathcal{O}(\hbar)\right]$$
$$+ \sum_{\boldsymbol{m}\in\mathbb{Z}^2\setminus\{0\}} \frac{|D|}{(2\pi\hbar)^{3/2}}\frac{\chi_s(\alpha_{\boldsymbol{m}})}{\sqrt{l_{\boldsymbol{m}}\sqrt{E}}}\cos\left(\frac{\sqrt{E}}{\hbar}l_{\boldsymbol{m}} - \frac{\pi}{4}\right)\left[1 + \mathcal{O}(\hbar)\right]. \tag{3.150}$$

On the other hand the Pauli equation

$$\hat{H}_{\mathrm{P}}\Psi = E\Psi \tag{3.151}$$

with Hamiltonian (3.126), $2m = 1$, and periodic boundary conditions can be solved trivially, and one immediately obtains the eigenvalues

$$E_{\boldsymbol{n},m_s} = (2\pi\hbar)^2\left(\frac{n_1^2}{a_1^2} + \frac{n_2^2}{a_2^2}\right) + m_s\kappa\, 2\pi\hbar^2\sqrt{\frac{n_1^2}{a_1^2} + \frac{n_2^2}{a_2^2}}\,, \tag{3.152}$$

with $\boldsymbol{n} \in \mathbb{Z}^2$ and spin quantum number $m_s = -s, -s+1, \ldots, s$. We are now going to derive an exact trace formula for the spectral density

$$d(E) = \sum_{\boldsymbol{n}\in\mathbb{Z}^2}\sum_{m_s=-s}^{s} \delta(E - E_{\boldsymbol{n},m_s})\,. \tag{3.153}$$

This goal can be achieved by Poisson summation, see Appendix A,

$$d(E) = \sum_{\boldsymbol{m}\in\mathbb{Z}^2}\int_{\mathbb{R}^2}\sum_{m_s=-s}^{s} \delta(E - E_{\boldsymbol{n},m_s})\,\mathrm{e}^{2\pi i \boldsymbol{m}\boldsymbol{n}}\,\mathrm{d}^2 n\,. \tag{3.154}$$

Rescaling the integration variables to $z_j := n_j/a_j$ and defining

$$\boldsymbol{\mu} := \begin{pmatrix} a_1 m_1 \\ a_2 m_2 \end{pmatrix} \tag{3.155}$$

the spectral density reads

$$d(E) = |D| \sum_{\boldsymbol{m}\in\mathbb{Z}^2} \int_{\mathbb{R}^2} \sum_{m_s=-s}^{s} \delta(E - E_{\boldsymbol{n},m_s})\, \mathrm{e}^{2\pi\mathrm{i}\boldsymbol{\mu z}}\, \mathrm{d}^2 z\,, \tag{3.156}$$

where $|D| = a_1 a_2$. Since the eigenvalues

$$E_{\boldsymbol{n},m_s} = (2\pi\hbar)^2|\boldsymbol{z}|^2 + m_s\kappa\, 2\pi\hbar^2|\boldsymbol{z}| \tag{3.157}$$

depend only on $|\boldsymbol{z}|$ it is convenient to introduce polar coordinates $(|\boldsymbol{z}|, \phi)$ for $\boldsymbol{z}$,

$$d(E) = |D| \sum_{\boldsymbol{m}\in\mathbb{Z}^2} \int_0^{2\pi} \int_0^{\infty} \sum_{m_s=-s}^{s} \delta(E - E_{\boldsymbol{n},m_s}))\, \mathrm{e}^{2\pi\mathrm{i}|\boldsymbol{\mu}||\boldsymbol{z}|\cos\phi} |\boldsymbol{z}|\, \mathrm{d}|\boldsymbol{z}|\, \mathrm{d}\phi\,, \tag{3.158}$$

where we have chosen the z_1-axis parallel to $\boldsymbol{\mu}$. The $|\boldsymbol{z}|$-integration can be performed exactly and with the Bessel function

$$J_0(y) = \frac{1}{2\pi} \int_0^{2\pi} \mathrm{e}^{\mathrm{i}y\cos(\phi)}\, \mathrm{d}\phi \tag{3.159}$$

we obtain the exact trace formula $(E > 0)$

$$\begin{aligned} d(E) = \frac{\pi|D|}{(2\pi\hbar)^2} \sum_{\boldsymbol{m}\in\mathbb{Z}^2} \sum_{m_s=-s}^{s} &\left(1 + \frac{\hbar m_s\kappa}{\sqrt{4E + \hbar^2 m_s^2\kappa^2}}\right) \\ &\times J_0\left(\frac{|\boldsymbol{\mu}|\sqrt{E}}{\hbar}\sqrt{1 + \frac{m_s^2\kappa^2\hbar^2}{4E}} + \frac{m_s\kappa|\boldsymbol{\mu}|}{2}\right) \end{aligned} \tag{3.160}$$

Before we compare with the semiclassical version (3.149) of the trace formula notice that $|\boldsymbol{\mu}| = l_{\boldsymbol{m}}$ as defined below (3.143). In order to determine the leading terms of (3.160) as $\hbar \to 0$ we employ the following asymptotics. For the factor on the first line of (3.160) we simply have

$$1 + \frac{\hbar m_s\kappa}{\sqrt{4E + \hbar^2 m_s^2\kappa^2}} = 1 + \mathcal{O}(\hbar)\,. \tag{3.161}$$

Using the asymptotic expansion of the Bessel function, see e.g. [46], the term on the second line of (3.160) becomes

$$\begin{aligned} &J_0\left(\frac{|\boldsymbol{\mu}|\sqrt{E}}{\hbar}\sqrt{1 + \frac{m_s^2\kappa^2\hbar^2}{4E}} + \frac{m_s\alpha|\boldsymbol{\mu}|}{2}\right) \\ &= J_0\left(\frac{l_{\boldsymbol{m}}\sqrt{E}}{\hbar} + \frac{m_s\kappa l_{\boldsymbol{m}}}{2}\right)[1 + \mathcal{O}(\hbar)] \\ &= \sqrt{\frac{2\hbar}{\pi l_{\boldsymbol{m}}\sqrt{E}}} \cos\left(\frac{l_{\boldsymbol{m}}\sqrt{E}}{\hbar} + \frac{m_s\kappa l_{\boldsymbol{m}}}{2} - \frac{\pi}{4}\right)[1 + \mathcal{O}(\hbar)]\,, \end{aligned} \tag{3.162}$$

Finally, due the addition formula of the cosine,

$$\cos\left(\frac{l_{\boldsymbol{m}}\sqrt{E}}{\hbar}+\frac{m_s\kappa l_{\boldsymbol{m}}}{2}-\frac{\pi}{4}\right)=\cos\left(\frac{l_{\boldsymbol{m}}\sqrt{E}}{\hbar}-\frac{\pi}{4}\right)\cos\left(\frac{m_s\kappa l_{\boldsymbol{m}}}{2}\right)$$
$$-\sin\left(\frac{l_{\boldsymbol{m}}\sqrt{E}}{\hbar}-\frac{\pi}{4}\right)\sin\left(\frac{m_s\kappa l_{\boldsymbol{m}}}{2}\right), \tag{3.163}$$

and since we have to sum over all spin projections $m_s = -s, \dots, s$, we have the further simplification

$$\sum_{m_s=-s}^{s}\cos\left(\frac{l_{\boldsymbol{m}}\sqrt{E}}{\hbar}+\frac{m_s\kappa l_{\boldsymbol{m}}}{2}-\frac{\pi}{4}\right)=\chi_s(\alpha_{\boldsymbol{m}})\cos\left(\frac{l_{\boldsymbol{m}}\sqrt{E}}{\hbar}-\frac{\pi}{4}\right), \tag{3.164}$$

where $\alpha_{\boldsymbol{m}}$ was defined in (3.147). Altogether, the semiclassically leading terms of the exact trace formula (3.160) are given by

$$d(E)=(2s+1)\,\frac{\pi|D|}{(2\pi\hbar)^2}\,\left[1+\mathcal{O}(\hbar)\right]$$
$$+\frac{|D|}{(2\pi\hbar)^{3/2}}\sum_{\boldsymbol{m}\in\mathbb{Z}^2\setminus\{0\}}\frac{\chi_s(\alpha_{\boldsymbol{m}})}{\sqrt{l_{\boldsymbol{m}}\sqrt{E}}}\cos\left(\frac{l_{\boldsymbol{m}}\sqrt{E}}{\hbar}-\frac{\pi}{4}\right)\left[1+\mathcal{O}(\hbar)\right]. \tag{3.165}$$

This result is – as it should be – identical to (3.150) which was obtained purely from classical dynamics.

3.6.4 Spin–Orbit Coupling in Non-Relativistic Hydrogen

We briefly want to discuss the issue of spin–orbit coupling in the hydrogen atom in a non-relativistic context. To this end one investigates the Pauli Hamiltonian

$$\hat{H}=-\frac{\hbar^2}{2m}\Delta-\frac{e}{|\boldsymbol{x}|}+\frac{e^2\hbar}{(2mc)^2}\frac{1}{|\boldsymbol{x}|^3}\boldsymbol{\sigma}\hat{\boldsymbol{L}}\,. \tag{3.166}$$

This Hamiltonian can either be obtained in a semi-heuristic way by considering the interaction energy of the magnetic moment of spin and the magnetic moment generated by a moving charge (the electron), as is demonstrated in most quantum mechanics text books, or by a non-relativistic approximation of the respective Dirac equation, see e.g [28, 29, 41]. A good estimate for the spin–orbit splitting of the levels of the spinless, non-relativistic hydrogen atom is then obtained by treating the last term of (3.166) in first order perturbation theory.

However, it can be proven that the Hamiltonian (3.166) as it stands has a purely continuous spectrum, see [47–50] and references therein. Nevertheless, it has been shown that first order perturbation theory for the above

Hamiltonian can be used in order to calculate some relativistic corrections to the eigenvalues of the non-relativistic hydrogen atom. We will not go into the details of this problem but simply emphasize that any theory based upon (3.166) has to be treated with caution. Thus, we will simply start from the well-known spectrum of the Hamiltonian (3.166) without spin–orbit coupling, i.e. from the Schrödinger equation of the hydrogen atom, which by Poisson summation gives rise to an exact trace formula. Then we show how to modify this trace formula along the lines discussed in the preceeding sections, in order to account for the spin contribution. The resulting formula is compatible with the first order perturbation theory, as we will see in Sect. 5.4.4, where we approach the problem in a slightly different way.

The energy levels of the non-relativistic hydrogen atom without spin are given by the Rydberg formula,

$$E_n = -\frac{R}{n^2}\,, \tag{3.167}$$

where $n \in \mathbb{N}$ and $R \approx 13.6\,\mathrm{eV}$ is the Rydberg energy. For each n we have eigenstates with angular momentum quantum numbers $l = 0, 1, \ldots, n-1$ and $|m| \leq l$ which are all degenerate. Thus, the spectral density reads

$$d(E) = \sum_{n=1}^{\infty}\sum_{l=0}^{n-1}\sum_{m=-l}^{l} \delta(E - E_n)\,. \tag{3.168}$$

Note that, since we are dealing with bound states, we always assume $E < 0$. Keeping in mind the spin–orbit interaction which is proportional to angular momentum, we first rearrange the sums and then apply the Poisson summation formula,

$$\begin{aligned} d(E) &= \sum_{l=0}^{\infty}\sum_{n=l+1}^{\infty} (2l+1)\delta(E - E_n) \\ &= \sum_{l=0}^{\infty}(2l+1)\sum_{k\in\mathbb{Z}}\int_{l+1}^{\infty} \delta\left(E + \frac{R}{n^2}\right) \mathrm{e}^{2\pi i k n}\,\mathrm{d}n \\ &= \sum_{l=0}^{\left[\sqrt{-\frac{R}{E}}\right]-1} (2l+1)\sum_{k\in\mathbb{Z}} \frac{\sqrt{R}}{2}(-E)^{-3/2}\,\mathrm{e}^{2\pi i k\sqrt{-R/E}}\,, \end{aligned} \tag{3.169}$$

where $[x]$ is the largest integer less than or equal to x. It can be easily shown that the exponent of the last term is the action of the k-fold repetition of a Kepler orbit multiplied by $(\mathrm{i}/\hbar)$. Thus, (3.169) is indeed a trace formula in the usual sense given by a sum over repetitions k of periodic orbits with angular momentum labeled by l.

In order to incorporate spin into this formula we simply have to insert a term $\chi_{1/2}(\alpha)$ where α is the spin rotation angle of spin precession along a

Kepler orbit with a certain angular momentum $\boldsymbol{L}$. Since angular momentum is constant along a Kepler ellipse the equation of spin precession,

$$\dot{\boldsymbol{s}} = \frac{e^2}{2m^2c^2}\frac{1}{|\boldsymbol{x}|^3}\boldsymbol{L} \times \boldsymbol{s} \,, \tag{3.170}$$

can be integrated directly. One finds that after a time t an initial spin vector $\boldsymbol{s}$ has been rotated about $\boldsymbol{L}$ by an angle

$$\alpha(t) = \frac{e}{2m^2c^2}|\boldsymbol{L}|\int_0^t \frac{\mathrm{d}t'}{|\boldsymbol{x}(t')|} \,. \tag{3.171}$$

We have to evaluate this angle at times $t = kT$, where T denotes the period of a Kepler orbit. The integral can be calculated introducing polar coordinates (r, ϕ) in the plane perpendicular to $\boldsymbol{L}$,

$$\int_0^{kT} \frac{\mathrm{d}t}{|\boldsymbol{x}(t)|^3} = \int_0^{2\pi k} \frac{\mathrm{d}\phi}{\dot{\phi}r^3} \,. \tag{3.172}$$

In polar coordinates the angular momentum is $|\boldsymbol{L}| = mr^2\dot{\phi}$ and the remaining integral is evaluated using the well-known formula $r(\phi)$ for a Kepler ellipse (see e.g. [35]), yielding

$$\int_0^{kT} \frac{\mathrm{d}t}{|\boldsymbol{x}(t)|^3} = \frac{m}{|\boldsymbol{L}|}\int_0^{2\pi k} \frac{\mathrm{d}\phi}{r} = \frac{2\pi k m^2 e^2}{|\boldsymbol{L}|^3} \,. \tag{3.173}$$

Thus, the weight factor to be inserted into (3.169) is given by

$$\chi_{1/2}(\alpha(kT)) = 2\cos\left(\frac{\pi e^4}{2c^2|\boldsymbol{L}|^2}k\right) \,. \tag{3.174}$$

The value of $|\boldsymbol{L}|^2$ should be taken to be $\hbar^2(l+1/2)^2$ rather than $\hbar^2 l(l+1)$. This is immediately clear if one treats the Schrödinger equation for the hydrogen atom on a semiclassical footing. However, if one first separates in spherical coordinates and then solves the radial equation by means of the WKB-method one first has $|\boldsymbol{L}|^2 = \hbar^2 l(l+1)$ which afterwards has to be corrected to $|\boldsymbol{L}|^2 = \hbar^2(l+1/2)^2$. A justification for this procedure was given by Langer [51]. Upon replacing $|\boldsymbol{L}|$ by $\hbar(l+1/2)$, the semiclassical trace formula for (3.166) reads

$$d(E) \sim \sum_{l=0}^{\left[\sqrt{-R/E}\right]-1} \sum_{k\in\mathbb{Z}} (2l+1)\cos\left(\frac{\pi e^4}{2c^2\hbar^2(l+\frac{1}{2})^2}k\right) \times \sqrt{R}\,(-E)^{-3/2}\,\mathrm{e}^{2\pi\mathrm{i}k\sqrt{-R/E}} \,. \tag{3.175}$$

Introducing Sommerfeld's fine structure constant $\alpha_S := e^2/(\hbar c)$ the result can be written as

$$d(E) \sim \sum_{l=0}^{\left[\sqrt{-R/E}\right]-1} \sum_{k\in\mathbb{Z}} (2l+1)\cos\left(\frac{\pi}{2}\frac{\alpha_S^2}{(l+\frac{1}{2})^2}k\right) \times \sqrt{R}\,(-E)^{-3/2}\,\mathrm{e}^{2\pi \mathrm{i} k\sqrt{-R/E}} \,. \tag{3.176}$$

We are now in the fortunate position that we are able to perform the k-sum exactly by Poisson summation. This gives an explicit formula for the eigenvalues as follows (cf. Appendix A),

$$d(E) \sim \sum_{l=0}^{\left[\sqrt{-R/E}\right]-1} \frac{2l+1}{2} \sum_{n\in\mathbb{Z}} \left[\delta\left(\sqrt{-\frac{R}{E}} + \frac{\alpha_S^2}{(2l+1)^2} - n\right) + \delta\left(\sqrt{-\frac{R}{E}} - \frac{\alpha_S^2}{(2l+1)^2} - n\right)\right], \tag{3.177}$$

i.e. we find δ-peaks at energies

$$E_{nl}^{\pm} := \frac{-R}{\left(n \pm \left(\frac{\alpha_S}{2l+1}\right)^2\right)^2} \,. \tag{3.178}$$

One might be surprised to see a sum over all integers n in (3.177), however, the l-sum still requires $\sqrt{-R/E} \geq l+1$. On the other hand from the δ-functions we have $\sqrt{-R/E} = n \pm \alpha_S^2/(2l+1)^2$, yielding the following relation between the quantum numbers n and l,

$$n - l - 1 \geq \mp \left(\frac{\alpha_S}{2l+1}\right)^2 \,. \tag{3.179}$$

Notice that since $l \geq 0$ we have $0 \leq \alpha_S/(2l+1) \leq \alpha_S$, i.e. if we restrict α_S to the interval $(0,1)$, which includes the physical value $\alpha_S = \frac{1}{137}$, we find the conditions

$$l \leq n-1 \quad \text{for} \quad E_{nl}^{+} \quad \text{and} \quad l \leq n-2 \quad \text{for} \quad E_{nl}^{-} \,, \tag{3.180}$$

which together with $l \geq 0$ also imply $n \geq 0$. Thus, from the semiclassical calculation we have obtained the trace formula

$$d(E) \sim \sum_{n=1}^{\infty}\sum_{l=0}^{n-1}(2l+1)\,\delta(E - E_{nl}^{+}) + \sum_{n=2}^{\infty}\sum_{l=0}^{n-2}(2l+1)\,\delta(E - E_{nl}^{-}) \,. \tag{3.181}$$

We postpone the discussion of this result to Sect. 5.4.4 where we will approach the same problem from a different direction.

3.7 Trace Formula for the Dirac Equation

So far we have only discussed semiclassics for the Pauli equation. In this section we want to show how the same concepts can be applied to the Dirac equation for relativistic particles with spin 1/2, mass m and charge e, see e.g. [28, 29, 41, 52]. The presentation in this section is a brief summary of [1–3] and for details we refer the reader to these works; the semiclassical time evolution kernel of the Dirac equation is also discussed in [53].

The Dirac equation can be written in the familiar form

$$\mathrm{i}\hbar \frac{\mathrm{d}\Psi}{\mathrm{d}t} = \hat{H}_{\mathrm{D}} \Psi \,, \tag{3.182}$$

where now $\Psi \in L^2(\mathbb{R}^3) \otimes \mathbb{C}^4$ is a four-spinor. The Dirac Hamiltonian $\hat{H}_{\mathrm{D}}$ for a particle in external electro-magnetic fields reads

$$\hat{H}_{\mathrm{D}} = c\boldsymbol{\alpha} \left(\frac{\hbar}{\mathrm{i}} \nabla - \frac{e}{c} \boldsymbol{A}(\boldsymbol{x}) \right) + e\phi(\boldsymbol{x})\, \mathbb{1}_4 + \beta mc^2 \,, \tag{3.183}$$

where $(\phi, \boldsymbol{A})$ are the electro-magnetic potentials and the 4×4 matrices $\boldsymbol{\alpha}$ and β can be chosen to be

$$\boldsymbol{\alpha} = \begin{pmatrix} 0 & \boldsymbol{\sigma} \\ \boldsymbol{\sigma} & 0 \end{pmatrix} \quad \text{and} \quad \beta = \begin{pmatrix} \mathbb{1}_2 & 0 \\ 0 & -\mathbb{1}_2 \end{pmatrix} . \tag{3.184}$$

In (3.183) we have indicated only a spatial dependence of the potentials $(\phi, \boldsymbol{A})$, since we are finally aiming at a semiclassical theory for bound state energy eigenvalues. Note, that the semiclassical reasoning immediately translates to the general case of potentials that depend on both $\boldsymbol{x}$ and t. With the methods presented here one can then still calculate a semiclassical time evolution kernel.

In the non-relativistic limit $c \to \infty$ the coupling between the upper and the lower pairs of components of Ψ becomes small and in leading order one finds the Pauli equation for a two-spinor consisting of the upper two components, see e.g. [28, 29]. For time independent electro-magnetic potentials which decay fast enough as $|\boldsymbol{x}| \to \infty$ and have no stronger singularities than Coulomb singularities the Dirac Hamiltonian has a continuous spectrum for energies $|E| > mc^2$ and a possible point spectrum with eigenvalues $E_n \in (-mc^2, mc^2)$, see [29] for details. Unoccupied states in the negative continuum can be interpreted as states of antiparticles of the electron, the positron. This interpretation was first given by Dirac himself in [54], where he suggested that these "holes" could describe protons. Realising that the particle described by such a hole necessarily has the same mass as the electron he predicted the existence of the positron [55], which shortly afterwards was found experimentally by Anderson [56, 57]. See also Dirac's Nobel lecture [58] for an overview.

The matrix valued Weyl symbol, cf. Appendix E, of (3.183),

$$H_{\mathrm{D}} = \begin{pmatrix} (e\phi + mc^2)\mathbb{1}_2 & c\boldsymbol{\sigma}\left(\frac{\hbar}{\mathrm{i}}\nabla - \frac{e}{c}\boldsymbol{A}\right) \\ c\boldsymbol{\sigma}\left(\frac{\hbar}{\mathrm{i}}\nabla - \frac{e}{c}\boldsymbol{A}\right) & (e\phi - mc^2)\mathbb{1}_2 \end{pmatrix}, \tag{3.185}$$

has two degenerate eigenvalues

$$H^{\pm}(\boldsymbol{p}, \boldsymbol{x}) = e\phi \pm \sqrt{c^2\left(\boldsymbol{p} - \frac{e}{c}\boldsymbol{A}\right)^2 + m^2c^4}\,. \tag{3.186}$$

If we interpret these as classical Hamiltonians, then H^- gives rise to dynamics with negative kinetic energy which should be associated with the antiparticles, cf. [2, 59]. The eigenvectors of H_{D} can be chosen as the columns of the 4×2 matrices

$$\begin{aligned} V_+(\boldsymbol{p}, \boldsymbol{x}) &= \frac{1}{\sqrt{2\epsilon(\epsilon + mc^2)}} \begin{pmatrix} (\epsilon + mc^2)\mathbb{1}_2 \\ c\boldsymbol{\sigma}\left(\boldsymbol{p} - \frac{e}{c}\boldsymbol{A}\right) \end{pmatrix} \quad \text{and} \\ V_-(\boldsymbol{p}, \boldsymbol{x}) &= \frac{1}{\sqrt{2\epsilon(\epsilon + mc^2)}} \begin{pmatrix} c\boldsymbol{\sigma}\left(\boldsymbol{p} - \frac{e}{c}\boldsymbol{A}\right) \\ -(\epsilon + mc^2)\mathbb{1}_2 \end{pmatrix}, \end{aligned} \tag{3.187}$$

i.e. $H_{\mathrm{D}}V_{\pm} = H^{\pm}V_{\pm}$, where we have used the abbreviation

$$\epsilon := \sqrt{c^2\left(\boldsymbol{p} - \frac{e}{c}\boldsymbol{A}\right)^2 + m^2c^4}\,. \tag{3.188}$$

With this choice the eigenvectors are orthogonal and normalised,

$$V_+^\dagger V_+ = \mathbb{1}_2 = V_-^\dagger V_-\,, \quad V_+^\dagger V_- = 0 = V_-^\dagger V_+\,, \tag{3.189}$$

and we have the projectors $\Pi_{\pm} = V_{\pm}V_{\pm}^\dagger$ with $\Pi_+ + \Pi_- = \mathbb{1}_4$.

As in the previously discussed case of the Pauli equation the main step in the derivation of a trace formula is finding a semiclassical expression for the time evolution kernel. Anticipating contributions with both positive and negative kinetic energies we modify the semiclassical ansatz (3.29) to

$$\begin{aligned} K(\boldsymbol{x}, \boldsymbol{y}, t) = \frac{1}{(2\pi\hbar)^3} \int_{\mathbb{R}^3} \Big[& a_\hbar^+(\boldsymbol{x}, \boldsymbol{\xi}, t)\, \mathrm{e}^{(\mathrm{i}/\hbar)(S^+(\boldsymbol{x},\boldsymbol{\xi},t) - \boldsymbol{y}\boldsymbol{\xi})} \\ & + a_\hbar^-(\boldsymbol{x}, \boldsymbol{\xi}, t)\, \mathrm{e}^{(\mathrm{i}/\hbar)(S^-(\boldsymbol{x},\boldsymbol{\xi},t) - \boldsymbol{y}\boldsymbol{\xi})} \Big]\, \mathrm{d}^3\xi \end{aligned} \tag{3.190}$$

with scalar functions $S^{\pm}$ and matrix valued amplitudes

$$a_\hbar^{\pm}(\boldsymbol{x}, \boldsymbol{\xi}, t) = \sum_{k \geq 0} \left(\frac{\hbar}{\mathrm{i}}\right)^k a_k^{\pm}(\boldsymbol{x}, \boldsymbol{\xi}, t)\,. \tag{3.191}$$

Upon inserting $K(\boldsymbol{x}, \boldsymbol{y}, t)$ into the Dirac equation we find in leading semiclassical order

$$\left[H_{\mathrm{D}}(\nabla_{\boldsymbol{x}} S^{+}, \boldsymbol{x}) + \frac{\partial S^{+}}{\partial t}\right] a_0^{+} + \left[H_{\mathrm{D}}(\nabla_{\boldsymbol{x}} S^{-}, \boldsymbol{x}) + \frac{\partial S^{-}}{\partial t}\right] a_0^{-} = 0\,. \tag{3.192}$$

This equation can be solved by choosing $S^{\pm}$ to fulfill the Hamilton–Jacobi equations

$$H^{\pm}(\nabla_{\boldsymbol{x}} S^{\pm}, \boldsymbol{x}) + \frac{\partial S^{\pm}}{\partial t} = 0 \tag{3.193}$$

and simultaneously letting the leading order term a_0 of the amplitudes be of the form

$$a_0^{\pm}(\boldsymbol{x}, \boldsymbol{\xi}, t) = V_{\pm}(\nabla_{\boldsymbol{x}} S^{\pm}, \boldsymbol{x})\, \tilde{V}_{\pm} \tag{3.194}$$

with some 2×4 matrices $\tilde{V}_{\pm}$ which still have to be determined. In next-to-leading order as $\hbar \to 0$ we have to solve

$$\begin{aligned} &\left[H_{\mathrm{D}}(\nabla_{\boldsymbol{x}} S^{+}, \boldsymbol{x}) + \frac{\partial S^{+}}{\partial t}\right] a_1^{+} + \left(c\boldsymbol{\alpha}\nabla_{\boldsymbol{x}} + \frac{\partial}{\partial t}\right) a_0^{+} \\ &+ \left[H_{\mathrm{D}}(\nabla_{\boldsymbol{x}} S^{-}, \boldsymbol{x}) + \frac{\partial S^{-}}{\partial t}\right] a_1^{-} + \left(c\boldsymbol{\alpha}\nabla_{\boldsymbol{x}} + \frac{\partial}{\partial t}\right) a_0^{-} = 0 \end{aligned} \tag{3.195}$$

Multiplying this equation from the left with $V_{+}^{\dagger}(\nabla_{\boldsymbol{x}} S^{\pm}, \boldsymbol{x})$ and $V_{-}^{\dagger}(\nabla_{\boldsymbol{x}} S^{\pm}, \boldsymbol{x})$, respectively, and using (3.189) we obtain the following equations for $\tilde{V}_{\pm}$,

$$V_{\pm}^{\dagger}(\nabla_{\boldsymbol{x}} S^{\pm}, \boldsymbol{x}) \left(c\boldsymbol{\alpha}\nabla_{\boldsymbol{x}} + \frac{\partial}{\partial t}\right) \left(V_{\pm}(\nabla_{\boldsymbol{x}} S^{\pm}, \boldsymbol{x})\tilde{V}_{\pm}\right) = 0\,. \tag{3.196}$$

For $\tilde{V}_{\pm}$ we may try the ansatz $\tilde{V}_{\pm} = b_{\pm} V_{\pm}^{\dagger}(\boldsymbol{\xi}, \nabla_{\boldsymbol{\xi}} S^{\pm})$ with 2×2 matrices $b_{\pm}$ which in terms of

$$a_0^{\pm}(\boldsymbol{x}, \boldsymbol{\xi}, t) = V_{\pm}(\nabla_{\boldsymbol{x}} S^{\pm}, \boldsymbol{x})\, b_{\pm}\, V_{\pm}^{\dagger}(\boldsymbol{\xi}, \nabla_{\boldsymbol{\xi}} S^{\pm}) \tag{3.197}$$

should be interpreted as follows. Acting with $K(\boldsymbol{x}, \boldsymbol{y}, t)$ on an initial spinor $\Psi(\boldsymbol{y})$ the matrix $V_{\pm}^{\dagger}(\boldsymbol{\xi}, \nabla_{\boldsymbol{\xi}} S^{\pm})$ projects onto the space of positive or negative kinetic energies, respectively, and maps the result to a two-spinor representation. Then follows propagation with $b_{\pm}$ up to time t. Finally, $V_{\pm}(\nabla_{\boldsymbol{x}} S^{\pm}, \boldsymbol{x})$ maps back to a four-spinor representation. After a tedious but straightforward calculation [2, 3] the transport equations for $b_{\pm}$ read

$$\begin{aligned} &\left[\frac{1}{2}\nabla_{\boldsymbol{x}}\left(\nabla_{\boldsymbol{p}} H^{\pm}(\nabla_{\boldsymbol{x}} S^{\pm}, \boldsymbol{x})\right) + \frac{\mathrm{i}}{2}\boldsymbol{\mathcal{B}}^{\pm}(\nabla_{\boldsymbol{x}} S^{\pm}, \boldsymbol{x})\,\boldsymbol{\sigma}\right] b_{\pm} \\ &+ \underbrace{\left[\left(\nabla_{\boldsymbol{p}} H^{\pm}(\nabla_{\boldsymbol{x}} S^{\pm}, \boldsymbol{x})\right)\nabla_{\boldsymbol{x}} + \frac{\partial}{\partial t}\right] b_{\pm}}_{=:\dot{b}_{\pm}} = 0 \end{aligned} \tag{3.198}$$

with

$$\boldsymbol{\mathcal{B}}^{\pm}(\boldsymbol{p}, \boldsymbol{x}) = \frac{ec^2}{\epsilon(\epsilon + mc^2)}\left[\left(\boldsymbol{p} - \frac{e}{c}\boldsymbol{A}\right) \times \boldsymbol{E}\right] \mp \frac{ec}{\epsilon}\boldsymbol{B}\,. \tag{3.199}$$

Here we have used the electric and magnetic fields

$$\begin{aligned} \boldsymbol{E}(\boldsymbol{x}) &:= -\nabla\phi(\boldsymbol{x}) \quad \text{and} \\ \boldsymbol{B}(\boldsymbol{x}) &:= \nabla \times \boldsymbol{A}(\boldsymbol{x}) \,. \end{aligned} \tag{3.200}$$

Since (3.198) is of the same form as (3.40) we can immediately exploit all results of Sects. 3.3–3.5.3. The amplitudes $b_\pm$ are given by

$$b_\pm = \sqrt{\det \frac{\partial^2 S^\pm}{\partial x \partial \xi}}\, d_\pm \tag{3.201}$$

where the 2×2 matrices $d_\pm$ have to solve the spin transport equations

$$\dot{d}_\pm + \frac{\mathrm{i}}{2}\boldsymbol{\sigma}\boldsymbol{\mathcal{B}}_\pm\, d_\pm = 0\,, \quad d_\pm\big|_{t=0} = \mathbb{1}_2\,. \tag{3.202}$$

Thus we have obtained an expression for the time evolution kernel of the Dirac equation which we can use in order to derive a trace formula for energies $E \in (-mc^2, mc^2)$; see [3] for technical details on how this energy localisation is performed. In general one finds contributions from periodic orbits of both classical dynamics, generated by H^+ or H^-, respectively. Since on a periodic orbit γ we have $V_\pm(\nabla_{\boldsymbol{x}} S, \boldsymbol{x}) = V_\pm(\boldsymbol{\xi}, \nabla_{\boldsymbol{\xi}} S)$, the matrix traces read

$$\operatorname{tr} a_0^\pm(\boldsymbol{x}, \boldsymbol{\xi}, T_\gamma) = \operatorname{tr} d_{\pm,\gamma} = \chi_{\frac{1}{2}}(\alpha_\gamma^\pm) = 2\cos(\alpha_\gamma^\pm/2)\,, \tag{3.203}$$

where we have used cyclic permutation for the terms inside the trace operation. As previously $\alpha_\gamma^\pm$ denotes the rotation angle which can be calculated from the equation of spin precession,

$$\dot{\boldsymbol{s}} = \boldsymbol{\mathcal{B}}^\pm \times \boldsymbol{s}\,, \tag{3.204}$$

along the orbit. Notice that, with the special form of the effective field $\boldsymbol{\mathcal{B}}^+$ (3.199), we have derived the equation of Thomas precession for a classical point particle with spin [60–62], see also [63], from the Dirac equation. Trace formulae for the Dirac equation are now of exactly the same form as those for Pauli operators discussed in Sect. 3.5. Sums over periodic orbits γ have to be replaced by sums over periodic orbits $\gamma^\pm$ with both positive and negative kinetic energies. The amplitudes have to be calculated from the respective dynamics generated by the Hamiltonians (3.186). Wherever the spin contribution $\chi_s(\alpha_\gamma)$ appears it has to be replaced by $\chi_{1/2}(\alpha_\gamma^\pm) = 2\cos(\alpha_\gamma^\pm/2)$, see [3] for details.

3.7.1 Reprise: The Dirac Oscillator

We want to illustrate the results from the previous section by revisiting the case of the Dirac oscillator, which was introduced in Sect. (2.1). Although the techniques needed are exactly the same as for a Dirac particle in external electro-magnetic fields, some details of the calculation are slightly different. Therefore, we will explicitly perform the whole semiclassical calculation instead of merely inserting special results into a general formula.

The quantum Hamiltonian (2.29) of the Dirac oscillator possesses the Weyl symbol

$$H_D(p,x) = \begin{pmatrix} mc^2 & cp + imc\omega x \\ cp - imc\omega x & -mc^2 \end{pmatrix}, \tag{3.205}$$

which has the two eigenvalues

$$H^{\pm}(p,x) = \pm\sqrt{m^2c^4 + c^2p^2 + c^2m^2\omega^2x^2}\,. \tag{3.206}$$

With a semiclassical ansatz

$$\begin{aligned} K(x,y,t) = \frac{1}{2\pi\hbar}\int_{\mathbb{R}} \Big[&\left(a_0^+(x,\xi,t) + \mathcal{O}(\hbar)\right) \mathrm{e}^{(\mathrm{i}/\hbar)(S^+(x,\xi,t)-y\xi)} \\ &+ \left(a_0^-(x,\xi,t) + \mathcal{O}(\hbar)\right) \mathrm{e}^{(\mathrm{i}/\hbar)(S^-(x,\xi,t)-y\xi)}\Big]\,\mathrm{d}\xi \end{aligned} \tag{3.207}$$

we obtain the leading order equation

$$\left[H_D\left(\frac{\partial S^+}{\partial x}, x\right) + \frac{\partial S^+}{\partial x}\right] a_0^+ + \left[H_D\left(\frac{\partial S^-}{\partial x}, x\right) + \frac{\partial S^-}{\partial x}\right] a_0^- = 0 \tag{3.208}$$

in the usual way. These can again be solved by letting $S^{\pm}$ fulfill the Hamilton–Jacobi equations

$$H^{\pm}\left(\frac{\partial S^{\pm}}{\partial x}, x\right) + \frac{\partial S^{\pm}}{\partial t} = 0 \tag{3.209}$$

with the leading order amplitude $a_0^{\pm}$ simultaneously being of the form

$$a_0^{\pm}(x,\xi,t) = v_{\pm}\left(\frac{\partial S^{\pm}}{\partial x}, x\right) w_{\pm}(x,\xi,t)\,. \tag{3.210}$$

Here $v_{\pm}$ denotes the eigenvectors of H_D, i.e. $H_D v_{\pm} = H^{\pm} v_{\pm}$, and the row vectors $w_{\pm}$ still need to be determined. One easily checks that the eigenvectors are given by

$$\begin{aligned} v_+(p,x) &= \frac{1}{\sqrt{2\epsilon(\epsilon + mc^2)}} \begin{pmatrix} \epsilon + mc^2 \\ cp - imc\omega x \end{pmatrix} \quad \text{and} \\ v_-(p,x) &= \frac{1}{\sqrt{2\epsilon(\epsilon + mc^2)}} \begin{pmatrix} cp - imc\omega x \\ -(\epsilon + mc^2) \end{pmatrix}, \end{aligned} \tag{3.211}$$

respectively, where ϵ is defined to be

$$\epsilon := \sqrt{m^2c^2 + c^2p^2 + m^2c^2\omega^2x^2} \tag{3.212}$$

cf. (3.188). The analogous equation to (3.195) reads

$$\begin{aligned}&\left[H_D\left(\frac{\partial S^+}{\partial x},x\right)+\frac{\partial S^+}{\partial x}\right]a_1^+ + \left(c\sigma_x\frac{\partial}{\partial x}+\frac{\partial}{\partial t}\right)a_0^+\\&+\left[H_D\left(\frac{\partial S^-}{\partial x},x\right)+\frac{\partial S^-}{\partial x}\right]a_1^- + \left(c\sigma_x\frac{\partial}{\partial x}+\frac{\partial}{\partial t}\right)a_0^- = 0\end{aligned} \tag{3.213}$$

After multiplying with $v_\pm^\dagger(\partial S^\pm/\partial x, x)$ from the left and making the ansatz

$$w_\pm(x,\xi,t) = \sqrt{\frac{\partial^2 S^\pm}{\partial x \partial \xi}}\; d_\pm \; v_\pm^\dagger\left(\xi, \frac{\partial S^\pm}{\partial \xi}\right) , \tag{3.214}$$

cf. (3.197), we are left with the following equations for the scalar functions $d_\pm$,

$$\dot{d}_\pm \mp \mathrm{i}\frac{c^2 m\omega}{2\epsilon}\, d_\pm = 0\, , \quad d_\pm\big|_{t=0} = 1\, . \tag{3.215}$$

Although spin in the usual sense did not make an appearance in this model, the second term in (3.215) can be considered as a remnant of the spin contributions $\mathrm{i}\boldsymbol{\sigma}\boldsymbol{\mathcal{B}}_\pm d_\pm/2$ in (3.202). The action S of a primitive periodic orbit of the Hamiltonians (3.206) was already given in (2.40). We find the period by differentiation with respect to E,

$$T = \frac{\mathrm{d}S}{\mathrm{d}E} = \frac{2\pi E}{mc^2\omega}\, . \tag{3.216}$$

Integrating (3.215) up to times kT and realising that the stationary phase condition in the t-integral of (3.66) reduces to $E = \epsilon$ we find the weight factor

$$d_{\pm,k} = \mathrm{e}^{\pm \mathrm{i}k\pi} \tag{3.217}$$

which together with the Maslov contribution $\mathrm{e}^{-\mathrm{i}\pi k\mu/2}$, $\mu = 2$, yields one. Thus, we have explained the missing phase factor in the trace formula (2.34) in terms of a remnant of spin transport. By putting together all the results of this section in the usual way, the reader easily verifies that the semiclassical trace formula for the Dirac oscillator is identical to the exact trace formula which was derived in Sect. 2.1.

3.8 A Different Limit of the Pauli Equation

As a last example for semiclassical trace formulae for multicomponent wave equations we will discuss a different limit of the Dirac equation in which one obtains different polarised classical Hamiltonians corresponding to $2s+1$ directions of the spin vector relative to the external field. This method is closely related to the discussion of spin–orbit coupling by Littlejohn and Flynn [30] based on their semiclassical theory developed in [64, 65]. Frisk and Guhr postulated a trace formula for this situation [31], which was only later fully derived from the Pauli equation [3]. Amann and Brack [66–68] discussed applications of this method in the context of nuclear physics.

Consider again the Pauli Hamiltonian (3.2),

$$\hat{H}_{\mathrm{P}} = \hat{H}\mathbb{1}_{2s+1} + \frac{\hbar}{2}\mathrm{d}\pi_s(\boldsymbol{\sigma})\hat{\boldsymbol{\mathcal{B}}} \, . \tag{3.218}$$

In this section we absorb the $\hbar$ in front of the spin-dependent term into the field $\hat{\boldsymbol{\mathcal{B}}}$ and thus discuss the semiclassical limit $\hbar \to 0$ for the spectrum of the (modified) Pauli Operator

$$\hat{H}_{\mathrm{P}} = \hat{H}\mathbb{1}_{2s+1} + \frac{1}{2}\mathrm{d}\pi_s(\boldsymbol{\sigma})\hat{\boldsymbol{\mathcal{B}}} \, , \tag{3.219}$$

which we also denote by $\hat{H}_{\mathrm{P}}$. Now the symbol of $\hat{H}_{\mathrm{P}}$,

$$H_{\mathrm{P}}(\boldsymbol{p},\boldsymbol{x}) = H(\boldsymbol{p},\boldsymbol{x}) + \frac{1}{2}\mathrm{d}\pi_s(\boldsymbol{\sigma})\boldsymbol{\mathcal{B}}(\boldsymbol{p},\boldsymbol{x}) \, , \tag{3.220}$$

has eigenvalues

$$H_{m_s}(\boldsymbol{p},\boldsymbol{x}) := H(\boldsymbol{p},\boldsymbol{x})\mathbb{1}_{2s+1} + m_s|\boldsymbol{\mathcal{B}}(\boldsymbol{p},\boldsymbol{x})| \, , \quad m_s = -s,\ldots,s \, . \tag{3.221}$$

These are all different if $|\boldsymbol{\mathcal{B}}(\boldsymbol{p},\boldsymbol{x})| \neq 0$, which we assume in the following. The points in phase space where $|\boldsymbol{\mathcal{B}}(\boldsymbol{p},\boldsymbol{x})|$ vanishes are known as mode conversion points and require special attention. We only return to this problem in the discussion at the end of this section.

Having different eigenvalues of the symbol we now make a semiclassical ansatz for the time evolution kernel of the Pauli equation with Hamiltonian (3.219) which is of the same structure as for the Dirac equation, cf. (3.190),

$$K(\boldsymbol{x},\boldsymbol{y},t) = \frac{1}{(2\pi\hbar)^d}\int_{\mathbb{R}^d}\sum_{m_s=-s}^{s}\left(a_0^{m_s}(\boldsymbol{x},\boldsymbol{\xi},t) + \frac{\hbar}{\mathrm{i}}\,a_1^{m_s}(\boldsymbol{x},\boldsymbol{\xi},t) + \ldots\right) \times \exp\left(\frac{\mathrm{i}}{\hbar}(S^{m_s}(\boldsymbol{x},\boldsymbol{\xi},t) - \boldsymbol{y}\boldsymbol{\xi})\right)\,\mathrm{d}^d\xi \, , \tag{3.222}$$

with scalar functions S^{m_s} and matrix valued amplitudes $a_j^{m_s}$. Upon inserting this ansatz into the Pauli equation for K, in leading order we obtain the matrix equation

$$\sum_{m_s=-s}^{s}\left[H_{\mathrm{P}}(\nabla_{\boldsymbol{x}}S^{m_s},\boldsymbol{x}) + \frac{\partial S^{m_s}}{\partial t}\right]a_0^{m_s} = 0 \, , \tag{3.223}$$

which only has non-trivial solutions if

$$H_{m_s}(\nabla_{\boldsymbol{x}}S^{m_s},\boldsymbol{x}) + \frac{\partial S^{m_s}}{\partial t} = 0 \, . \tag{3.224}$$

If we denote by $v_{m_s}(\boldsymbol{p},\boldsymbol{x})$ the eigenvector of $H_{\mathrm{P}}(\boldsymbol{p},\boldsymbol{x})$ corresponding to the eigenvalue $H_{m_s}(\boldsymbol{p},\boldsymbol{x})$, in order to fulfill (3.223) we also have the requirement that $a_0^{m_s}$ is of the form

$$a_0^{m_s}(\boldsymbol{x},\boldsymbol{\xi},t) = v_{m_s}(\nabla_{\boldsymbol{x}} S,\boldsymbol{x})\, w_{m_s}(\boldsymbol{x},\boldsymbol{\xi},t) \ , \tag{3.225}$$

with a row vector w_{m_s}, similar to (3.194). Note that due to the special form of the Hamiltonian, an eigenvector v_{m_s} of H_{P} is also an eigenvector of the spin dependent part, i.e.

$$\frac{1}{2}\,\mathrm{d}\pi_s(\boldsymbol{\sigma})\boldsymbol{\mathcal{B}}(\boldsymbol{p},\boldsymbol{x})\, v_{m_s}(\boldsymbol{p},\boldsymbol{x}) = m_s|\boldsymbol{\mathcal{B}}(\boldsymbol{p},\boldsymbol{x})|\, v_{m_s}(\boldsymbol{p},\boldsymbol{x}) \ . \tag{3.226}$$

The so far undetermined row vector w_{m_s} can be obtained from the next-to-leading order equation, which we find when inserting the ansatz for K into the Pauli equation, see (E.19),

$$\sum_{m_s=-s}^{s} \left\{ \left[H_{\mathrm{P}}(\nabla_{\boldsymbol{x}} S^{m_s},\boldsymbol{x}) + \frac{\partial S^{m_s}}{\partial t} \right] a_1^{m_s} \right.$$
$$\left. + \left[\frac{\partial}{\partial t} + \nabla_{\boldsymbol{p}} H_{\mathrm{P}}(\nabla_{\boldsymbol{x}} S^{m_s},\boldsymbol{x})\,\nabla_{\boldsymbol{x}} + \frac{1}{2}\nabla_{\boldsymbol{x}}\left(\nabla_{\boldsymbol{p}} H_{\mathrm{P}}(\nabla_{\boldsymbol{x}} S^{m_s},\boldsymbol{x})\right) \right] a_0^{m_s} \right\} = 0 \ . \tag{3.227}$$

As in previously discussed cases, we multiply with $v_{m_s}^\dagger$ from the left yielding

$$v_{m_s}^\dagger \Bigg[\underbrace{\frac{\partial}{\partial t} + \nabla_{\boldsymbol{p}} H_{m_s}(\nabla_{\boldsymbol{x}} S^{m_s},\boldsymbol{x})\,\nabla_{\boldsymbol{x}}}_{:=\frac{\mathrm{d}}{\mathrm{d}t}} + \frac{1}{2}\nabla_{\boldsymbol{x}}\left(\nabla_{\boldsymbol{p}} H_{m_s}(\nabla_{\boldsymbol{x}} S^{m_s},\boldsymbol{x})\right) \Bigg] v_{m_s} w_{m_s} = 0 \ . \tag{3.228}$$

Again we can separate spin and translational contributions making the ansatz

$$w_{m_s}(\boldsymbol{x},\boldsymbol{\xi},t) = \sqrt{\det \frac{\partial^2 S^{m_s}}{\partial x \partial \xi}}\; \mathrm{e}^{\mathrm{i}\eta_{m_s}}\, v_{m_s}^\dagger(\boldsymbol{\xi},\nabla_{\boldsymbol{\xi}} S^{m_s}) \ . \tag{3.229}$$

Then, using the results of Appendix B, for the scalar phase η we find the equation

$$\frac{\mathrm{d}\eta_{m_s}}{\mathrm{d}t} = \mathrm{i}\, v_{m_s}^\dagger \frac{\mathrm{d}v_{m_s}}{\mathrm{d}t} \ . \tag{3.230}$$

This is the familiar equation from the standard example of Berry's phase for a precessing spin [69]. For a periodic orbit the solution is given by [69, eq. (27)]

$$\eta_{m_s} = -m_s \Omega \tag{3.231}$$

where Ω denotes the solid angle enclosed by the curve $\boldsymbol{\mathcal{B}}$ evaluated along the orbit. Mathematically the phase η is the flux of a magnetic monopole of strength $-m_s$ located at the origin in $\boldsymbol{\mathcal{B}}$-space through the solid angle Ω.

Following the same steps as in Sect. 3.5 we can write down a trace formula for the spectrum of the Pauli operator (3.219) in the limit $\hbar \to 0$. For instance in the case of only isolated and unstable periodic orbits we find

$$\begin{aligned}\sum_n \varrho\left(\frac{E_n - E}{\hbar}\right) &= \frac{\hat{\varrho}(0)}{2\pi} \sum_{m_s=-s}^{s} \frac{|\Omega_E^{m_s}|}{(2\pi\hbar)^{d-1}} \left[1 + \mathcal{O}(\hbar)\right] \\ &+ \sum_{m_s=-s}^{s} \sum_{\gamma_{m_s}} \frac{\hat{\varrho}(T_{\gamma_{m_s}})}{2\pi} \frac{T^{\#}_{\gamma_{m_s}} \, \mathrm{e}^{\mathrm{i}\eta_{\gamma_{m_s}}}}{|\det(\mathbb{M}_{\gamma_{m_s}} - \mathbb{1}_{2d-2})|^{1/2}} \\ &\qquad \exp\left(\frac{\mathrm{i}}{\hbar} S_{\gamma_{m_s}} - \mathrm{i}\frac{\pi}{2}\mu_{\gamma_{m_s}}\right) \left[1 + \mathcal{O}(\hbar)\right] ,\end{aligned} \tag{3.232}$$

The energy shells $\Omega_E^{m_s}$, periods $T_{\gamma_{m_s}}$, actions $S_{\gamma_{m_s}}$, monodromy matrices $\mathbb{M}_{\gamma_{m_s}}$ and Maslov indices $\mu_{\gamma_{m_s}}$ are defined as in Sect. 3.5.2 but now with the Hamiltonians (3.221).

In [3] it was also investigated how the two expressions (3.104) and (3.232) are related. For small couplings $|\boldsymbol{\mathcal{B}}|$ the orbits γ_{m_s} of the polarised Hamiltonians (3.221) approach the orbits γ of the classical translational Hamiltonian without spin, which appear in (3.104). Conversely, if the classical spin precession is adiabatic, the spin weights $\chi_s(\alpha_\gamma)$ give rise to the factors $\exp(-\mathrm{i}\eta_{m_s})$, see [3] for details. This correspondence is also briefly discussed in [43].

In the derivation of (3.232) we have assumed that the eigenvalues of the symbol H_P, the Hamiltonians H_{m_s}, are all distinct. However, if the field $\boldsymbol{\mathcal{B}}$ vanishes at some points in phase space, the formalism breaks down, since one also has to consider orbits moving from one energy shell Ω_{m_s} to a different one. This is known as the problem of mode conversion, which in the general case is unsolved. For particular model systems Frisk and Guhr successfully took mode conversion into account in a trace formula by an ad hoc rule [31]. In the one dimensional case a general solution for a mode conversion point involving two modes was given by Flynn and Littlejohn [70]. A numerical analysis showing that in general one needs both types of orbits, those staying on a particular energy shell and those crossing, can be found in [67, 68]. We do not want to go into details on this subject, since throughout the work we will focus on the formalism developed in Sects. 3.1–3.5. This formalism is also more fundamental in the sense that it yields the classical equations of motion for both the translational degrees of freedom and for the precessing spin. Also, as we have seen in Sect. 3.7, the same structure is revealed by the semiclassical analysis of the Dirac equation, the relativistic wave equation from which the Pauli equation can be derived in the non-relativistic limit. However, in practical applications, one is not necessarily in the asymptotic regime and the coupling term $(\hbar/2)\mathrm{d}\pi_s(\boldsymbol{\sigma})\hat{\boldsymbol{\mathcal{B}}}$ can be quite large. For instance in [66–68] applications to spin–orbit coupling in nuclear physics are discussed, and it is shown that, depending on the particular situation, both methods prove useful. We also briefly comment on this subject again in Sect. 7.4 when discussing a special example.

References

1. J. Bolte, S. Keppeler: Phys. Rev. Lett. **81**, 1987–1991 (1998)
2. S. Keppeler: (1998), "Semiklassik für Dirac-Teilchen", Diplomarbeit, Universität Ulm, 1998
3. J. Bolte, S. Keppeler: Ann. Phys. (NY) **274**, 125–162 (1999)
4. J. Bolte: Found. Phys. **31**, 423–444 (2000)
5. J. Bolte: "The Gutzwiller trace formula for quantum systems with spin", in *Advances in Solid States Physics*, ed. by B. Kramer, Vol. 41 (Springer-Verlag, 2001), pp. 447–458
6. M.C. Gutzwiller: J. Math. Phys. **8**, 1979–2000 (1967)
7. M.C. Gutzwiller: J. Math. Phys. **10**, 1004–1020 (1969)
8. M.C. Gutzwiller: J. Math. Phys. **11**, 1791–1806 (1970)
9. M.C. Gutzwiller: J. Math. Phys. **12**, 343–358 (1971)
10. R. Balian, C. Bloch: Ann. Phys. (NY) **69**, 76–160 (1972)
11. R. Balian, C. Bloch: Ann. Phys. (NY) **85**, 514–545 (1974)
12. Y. Colin de Verdière: Compositio Mathematica **27**, 83–106 (1973)
13. Y. Colin de Verdière: Compositio mathematica **27**, 159–184 (1973)
14. J. Chazarain: Invent. Math. **24**, 65–82 (1974)
15. J.J. Duistermaat, V.W. Guillemin: Inv. Math. **29**, 39–79 (1975)
16. L. Hörmander: *The Analysis of Linear Partial Differential Operators*, Vol. I–IV, 2nd edn. (Springer-Verlag, 1990)
17. G.B. Folland: *Harmonic Analysis in Phase Space*, no. 122 in Annals of Mathematics Studies (Princeton University Press, Princeton, 1989)
18. A. Grigis, J. Sjöstrand: *Microlocal Analysis for Differential Operators*, London Mathematical Society Lecture Note Series (Cambridge University Press, Cambridge, 1994)
19. E. Meinrenken: Rep. Math. Phys. **31**, 279–295 (1992)
20. T. Paul, A. Uribe: J. Funct. Anal. **132**, 192–249 (1995)
21. M.C. Gutzwiller: *Chaos in Classical and Quantum Mechanics* (Springer-Verlag, New York, 1990)
22. D. Robert: *Autour de l'Approximation Semi-Classique* (Birkhäuser, Boston, 1987)
23. A.M. Ozorio de Almeida: *Hamiltonian Systems: Chaos and Quantization* (Cambridge University Press, Cambridge, 1988)
24. M. Tabor: *Chaos and Integrability in Nonlinear Dynamics* (John Wiley & Sons Inc., New York, 1989), an introduction
25. M. Brack, R.K. Bhaduri: *Semiclassical Physics* (Addison-Wesley, Reading, Massachusetts, 1997)
26. R. Blümel, W.P. Reinhardt: *Chaos in Atomic Physics* (Cambridge University Press, Cambridge, 1997)
27. F. Haake: *Quantum Signatures of Chaos*, 2nd edn. (Springer-Verlag, Berlin Heidelberg, 2001)
28. J.D. Bjorken, S.D. Drell: *Relativistic Quantum Mechanics* (McGraw-Hill, New York, St. Louis, San Francisco, 1964)

29. B. Thaller: *The Dirac Equation* (Springer-Verlag, Berlin, Heidelberg, 1992)
30. R.G. Littlejohn, W.G. Flynn: Phys. Rev. A **45**, 7697–7717 (1992)
31. H. Frisk, T. Guhr: Ann. Phys. (NY) **221**, 229–257 (1993)
32. J.H. Van Vleck: Proc. Nat. Acad. Sci. U.S.A. **14**, 178–188 (1928)
33. P. Choquard, F. Steiner: Helv. Phys. Acta **69**, 636–654 (1996)
34. A. Messiah: *Quantum Mechanics. Vol. I*, Translated from the French by G. M. Temmer (North-Holland Publishing Co., Amsterdam, 1961)
35. H. Goldstein: *Classical Mechanics*, 2nd edn. (Addison-Wesley, Reading, Massachusetts, 1980)
36. V.I. Arnold: *Mathematical Methods of Classical Mechanics* (Springer-Verlag, New York, 1978)
37. P. Choquard: Helv. Phys. Acta **28**, 89–157 (1955)
38. S.C. Creagh, R.G. Littlejohn: Phys. Rev. A **44**, 836–850 (1991)
39. S.C. Creagh, J.M. Robbins, R.G. Littlejohn: Phys. Rev. A **42**, 1907–1922 (1990)
40. J.M. Robbins: Nonlinearity **4**, 343–361 (1991)
41. P. Strange: *Relativistic Quantum Mechanics with Applications in Condensed Matter and Atomic Physics* (Cambridge University Press, Cambridge, 1998)
42. Yu.A. Bychkov, E.I. Rashba: J. Phys. C **17**, 6039–6045 (1984)
43. S. Keppeler, R. Winkler: Phys. Rev. Lett. **88**, 046 401 (2002)
44. E.I. Rashba: Sov. Phys.-Solid State **2**, 1109–1122 (1960). Translated from Fizika Tverdogo Tela **2** (1960) 1224–1238
45. R. Winkler: *Spin-Orbit Coupling Effects in Two-Dimensional Electron and Hole Systems*, Springer Tracts in Modern Physics (Springer-Verlag, Berlin Heidelberg New York, in press)
46. M. Abramowitz, I.A. Stegun (Eds.): *Pocketbook of Mathematical Functions*, abridged edn. (Verlag Harri Deutsch, Thun – Frankfurt/Main, 1984)
47. F. Gesztesy, H. Grosse, B. Thaller: Phys. Lett. B **116**, 155–157 (1982)
48. F. Gesztesy, B. Thaller, H. Grosse: Phys. Rev. Lett. **50**, 625–628 (1983)
49. F. Gesztesy, H. Grosse, B. Thaller: Ann. Inst. H. Poincaré Phys. Théor. **40**, 159–174 (1984)
50. F. Gesztesy, H. Grosse, B. Thaller: Adv. in Appl. Math. **6**, 159–176 (1985)
51. R.E. Langer: Phys. Rev. **51**, 669–676 (1937)
52. P.A.M. Dirac: Proc. R. Soc. London Ser. A **117**, 610–624 (1928)
53. K. Yajima: J. Fac. Sci. Univ. Tokyo Sect. IA Math. **29**, 161–194 (1982)
54. P.A.M. Dirac: Proc. R. Soc. London Ser. A **126**, 360–365 (1930)
55. P.A.M. Dirac: Proc. Roy. Soc. **133**, 60–72 (1931)
56. C.D. Anderson: Science **76**, 238–239 (1932)
57. C.D. Anderson: Phys. Rev. **43**, 491–494 (1933)
58. P.A.M. Dirac: "Theory of electrons and positrons", in *Nobel Lectures, Physics 1922-1941* (Elsevier, 1933), pp. 320–325
59. W. Pauli: Helv. Phys. Acta **5**, 179–199 (1932)
60. L.H. Thomas: Nature **117**, 514 (1926)
61. L.H. Thomas: Philos. Mag. **3**, 1–22 (1927)
62. V. Bargman, L. Michel, V.L. Telegdi: Phys. Rev. Lett. **2**, 435–436 (1959)
63. J.D. Jackson: *Klassische Elektrodynamik*, 2nd edn. (Walter de Gruyter, Berlin, New York, 1982)
64. R.G. Littlejohn, W.G. Flynn: Phys. Rev. Lett. **66**, 2839–2842 (1991)
65. R.G. Littlejohn, W.G. Flynn: Phys. Rev. A **44**, 5239–5256 (1991)

66. M. Brack, C. Amann: "Semiclassical calculation of shell effects in deformed nuclei", in *Fission Dynamics of Atomic Clusters and Nuclei* (World Scientific Publishing, 2001), p. 5
67. C. Amann: (2001), "Semiklassische Näherungen zur Spin-Bahn Kopplung", PhD thesis, Universität Regensburg, 2001
68. C. Amann, M. Brack: J. Phys. A **35**, 6009–6032 (2002)
69. M.V. Berry: Proc. R. Soc. London Ser. A **392**, 45–57 (1984)
70. W.G. Flynn, R.G. Littlejohn: Ann. Physics (NY) **234**, 334–403 (1994)

4 Classical Dynamics of Spinning Particles – the Skew Product

In the previous chapter we have seen that semiclassical trace formulae with spin require as an input the periodic orbits of a Hamiltonian flow ϕ^t on the classical phase space of the translational degrees of freedom and in addition certain weight factors related to spin transport or spin precession along these orbits. The spin dynamics is described by a time dependent differential equation, either on the group SU(2),

$$\dot{d} + \frac{\mathrm{i}}{2} \boldsymbol{\sigma} \boldsymbol{\mathcal{B}}(\phi^t(\boldsymbol{p}, \boldsymbol{x}))\, d = 0 \ , \tag{4.1}$$

or on the two-sphere S^2,

$$\dot{\boldsymbol{s}} = \boldsymbol{\mathcal{B}}(\phi^t(\boldsymbol{p}, \boldsymbol{x})) \times \boldsymbol{s} \ . \tag{4.2}$$

The group element $d \in \mathrm{SU}(2)$ and the classical spin vector $\boldsymbol{s} \in S^2 \hookrightarrow \mathbb{R}^3$ are both functions of the initial point $(\boldsymbol{p}, \boldsymbol{x})$ in the phase space of the translational degrees of freedom, where one starts the integration of (4.2), and of the time t. Through the field $\boldsymbol{\mathcal{B}}$, which is evaluated along a trajectory of ϕ^t, the equation is "driven" by the translational dynamics.

Since we are heading for questions originating in the field of quantum chaos – characterising properties of a quantum system by the degree of chaoticity or regularity, respectively, of the corresponding classical system – two problems arise. Firstly, we have to deal with two dynamics rather than one, which, furthermore, are linked in one direction only, the translational dynamics influencing the spin dynamics but not the other way round. We would rather be dealing with only one classical counterpart instead, a dynamical system the chaotic properties of which influence the quantum mechanics. Secondly the spin dynamics is non-autonomous, but the properties we intend to talk about – e.g. integrability or ergodicity – are most conveniently dealt with in the language of flows. Therefore, we will now recombine spin and translational dynamics, which were separated in the semiclassical treatment, into one flow.

4.1 The Skew Products Y^t and Y^t_{cl}

We can construct a flow for both translational and spin degrees of freedom by joining the Hamiltonian flow ϕ^t and the solution $d(\boldsymbol{p}, \boldsymbol{x}, t)$ of the spin transport equation (4.1) as follows,

$$\begin{aligned} Y^t : \mathbb{R}^d \times \mathbb{R}^d \times \mathrm{SU}(2) &\to \mathbb{R}^d \times \mathbb{R}^d \times \mathrm{SU}(2) \\ (\boldsymbol{p}, \boldsymbol{x}, g) &\mapsto (\phi^t(\boldsymbol{p}, \boldsymbol{x}), d(\boldsymbol{p}, \boldsymbol{x}, t)g) \; . \end{aligned} \tag{4.3}$$

In ergodic theory this kind of product flow is called a skew product, see e.g. [1, 2]. Since the dynamics "added" to the flow ϕ^t lives on a group this skew product is also called a group extension, or more precisely an SU(2)-extension of ϕ^t. First notice that due to the composition law (3.48) for the spin transporter d we have

$$Y^{t'+t} = Y^{t'} \circ Y^t \quad (\text{and} \quad Y^0 = id) \; , \tag{4.4}$$

i.e. Y^t is indeed a flow. The vanishing back reaction of the spin dynamics on the translational degrees of freedom can be expressed as follows: Defining the projection

$$\begin{aligned} \pi : \mathbb{R}^d \times \mathbb{R}^d \times \mathrm{SU}(2) &\to \mathbb{R}^d \times \mathbb{R}^d \\ (\boldsymbol{p}, \boldsymbol{x}, g) &\mapsto (\boldsymbol{p}, \boldsymbol{x}) \end{aligned} \tag{4.5}$$

we have

$$\pi \circ Y^t = \phi^t \circ \pi \; . \tag{4.6}$$

The skew product flow Y^t is also measure preserving. In order to see this recall that the translational dynamics ϕ^t is a Hamiltonian flow. Thus for an autonomous system its restriction to the energy shell

$$\Omega_E := \{(\boldsymbol{p}, \boldsymbol{x}) \,|\, H(\boldsymbol{p}, \boldsymbol{x}) = E\} \; , \tag{4.7}$$

which we shall assume to be compact, leaves Liouville measure μ_E,

$$\mathrm{d}\mu_E := \frac{\delta(H(\boldsymbol{p}, \boldsymbol{x}) - E)}{|\Omega_E|} \, \mathrm{d}^d p \, \mathrm{d}^d x \; , \tag{4.8}$$

invariant. On the group SU(2) we denote by μ_H the unique (normalised) left and right invariant Haar measure. Since Y^t acts on SU(2) by left multiplication with group elements $d \in \mathrm{SU}(2)$, the restriction of Y^t to $\Omega_E \times \mathrm{SU}(2)$ leaves the measure

$$\mu := \mu_E \times \mu_H \tag{4.9}$$

invariant. From time to time we will find it convenient to make use of the group structure, to which end we defined the flow Y^t. However, since the spin dynamics enters through the spin transporter $d \in \mathrm{SU}(2)$ this is a mixed representation with the translational degrees of freedom being classical but the spin dynamics still being treated on a quantum mechanical level.

Motivated by this asymmetry we also introduce a purely classical skew product Y^t_{cl} where the spin degrees of freedom are represented by a vector

$\boldsymbol{s}$ of constant length, i.e. the spin now lives on S^2. This can be achieved by defining

$$\begin{aligned} Y^t_{\rm cl} : \mathbb{R}^d \times \mathbb{R}^d \times S^2 &\rightarrow \mathbb{R}^d \times \mathbb{R}^d \times S^2 \\ (\boldsymbol{p}, \boldsymbol{x}, \boldsymbol{s}) &\mapsto (\phi^t(\boldsymbol{p}, \boldsymbol{x}), R(\boldsymbol{p}, \boldsymbol{x}, t)\boldsymbol{s}) , \end{aligned} \tag{4.10}$$

where $R(\boldsymbol{p}, \boldsymbol{x}, t) = \varphi(d(\boldsymbol{p}, \boldsymbol{x}, t))$ was defined in (3.55), i.e. $\boldsymbol{s}(t) := R(\boldsymbol{p}, \boldsymbol{x}, t)\boldsymbol{s}(0)$ solves (4.2). Let us first check that $Y^t_{\rm cl}$ is indeed a flow. As for the half classical skew product Y^t defined above the composition law

$$Y^{t'+t}_{\rm cl} = Y^{t'}_{\rm cl} \circ Y^t_{\rm cl} \tag{4.11}$$

holds provided that we have the property

$$R(\boldsymbol{p}, \boldsymbol{x}, t + t') = R(\phi^t(\boldsymbol{p}, \boldsymbol{x}), t')\, R(\boldsymbol{p}, \boldsymbol{x}, t) . \tag{4.12}$$

The latter can be derived from (3.48) and the definition (3.55) as follows.

$$\begin{aligned} R(\boldsymbol{p}, \boldsymbol{x}, t + t') &= \varphi\,(d(\boldsymbol{p}, \boldsymbol{x}, t + t')) \\ &= \varphi\left(d(\phi^t_H(\boldsymbol{p}, \boldsymbol{x}), t')\, d(\boldsymbol{p}, \boldsymbol{x}, t)\right) \\ &\underset{(*)}{=} \varphi\left(d(\phi^t_H(\boldsymbol{p}, \boldsymbol{x}), t')\right)\, \varphi\,(d(\boldsymbol{p}, \boldsymbol{x}, t)) \\ &= R(\phi^t(\boldsymbol{p}, \boldsymbol{x}), t')\, R(\boldsymbol{p}, \boldsymbol{x}, t) \end{aligned} \tag{4.13}$$

where in $(*)$ we used that the adjoint map $\mathrm{Ad}_g : \mathfrak{su}(2) \ni X \mapsto gXg^{-1} \in \mathfrak{su}(2)$ is indeed a representation of the group on its own Lie algebra, i.e. $\mathrm{Ad}_{gh} = \mathrm{Ad}_g \circ \mathrm{Ad}_h$ and thus $\varphi(gh) = \varphi(g)\,\varphi(h)$, cf. Appendix C. Since $Y^t_{\rm cl}$ is a skew product the translational dynamics is not influenced by the spin dynamics, i.e. using the (natural) projection

$$\begin{aligned} \pi_{\rm cl} : \mathbb{R}^d \times \mathbb{R}^d \times S^2 &\rightarrow \mathbb{R}^d \times \mathbb{R}^d \\ (\boldsymbol{p}, \boldsymbol{x}, \boldsymbol{s}) &\mapsto (\boldsymbol{p}, \boldsymbol{x}) , \end{aligned} \tag{4.14}$$

we have $\pi_{\rm cl} \circ Y^t_{\rm cl} = \phi^t \circ \pi_{\rm cl}$. This can also be illustrated by the commuting diagram

$$\begin{array}{ccc} (\boldsymbol{p}, \boldsymbol{x}, \boldsymbol{s}) & \xrightarrow{\quad Y^t_{\rm cl} \quad} & (\phi^t(\boldsymbol{p}, \boldsymbol{x}), R(\boldsymbol{p}, \boldsymbol{x}, t)\boldsymbol{s}) \\ \downarrow \pi_{\rm cl} & & \downarrow \pi_{\rm cl} \\ (\boldsymbol{p}, \boldsymbol{x}) & \xrightarrow{\quad \phi^t \quad} & \phi^t(\boldsymbol{p}, \boldsymbol{x}) \end{array} .$$

The spin dynamics, given by rotations $\varphi(d(\boldsymbol{p}, \boldsymbol{x}, \boldsymbol{s})) \in \mathrm{SO}(3)$, preserves the area on S^2 measured by μ_{S^2}, which in spherical coordinates,

$$s =: \begin{pmatrix} \sin\theta\cos\phi \\ \sin\theta\sin\phi \\ \cos\theta \end{pmatrix} , \tag{4.15}$$

reads

$$\mathrm{d}\mu_{S^2}(s) = \frac{1}{4\pi} \sin\theta \, \mathrm{d}\theta \, \mathrm{d}\phi \, . \tag{4.16}$$

Thus we conclude that the skew product Y_{cl}^t on $\mathbb{R}^d \times \mathbb{R}^d \times S^2$ preserves the product measure $\mu_{\mathrm{cl}} := \mu_E \times \mu_{S^2}$. The spin dynamics can be rephrased stressing that S^2 endowed with the metric $\mathrm{d}\mu_{S^2}$ is a symplectic manifold. For instance we can choose

$$p_{\mathrm{S}} = \cos\theta \quad \text{and} \quad x_{\mathrm{S}} = \phi \tag{4.17}$$

as canonically conjugate coordinates. Then the symplectic two-form $\mathrm{d}p_{\mathrm{S}} \wedge \mathrm{d}x_{\mathrm{S}}$ corresponds – up to normalisation – to the measure μ_{S^2} and the spin dynamics on S^2 are given by Hamilton's equations of motion,

$$\dot{p_{\mathrm{S}}} = -\frac{\partial H_{\mathrm{S}}}{\partial x_{\mathrm{S}}} \quad \text{and} \quad \dot{x_{\mathrm{S}}} = \frac{\partial H_{\mathrm{S}}}{\partial p_{\mathrm{S}}} , \tag{4.18}$$

with Hamiltonian

$$H_{\mathrm{S}} = \boldsymbol{\mathcal{B}} s \, , \tag{4.19}$$

which can be checked as follows: With (4.17) the classical spin vector reads

$$s = \begin{pmatrix} \sqrt{1-p_{\mathrm{S}}^2}\cos x_{\mathrm{S}} \\ \sqrt{1-p_{\mathrm{S}}^2}\sin x_{\mathrm{S}} \\ p_{\mathrm{S}} \end{pmatrix} \tag{4.20}$$

and thus we have

$$\begin{aligned} \dot{s}_z = \dot{p}_{\mathrm{S}} &= -\sqrt{1-p_{\mathrm{S}}^2}\, \left(\mathcal{B}_x(-\sin x_{\mathrm{S}}) + \mathcal{B}_y \cos x_{\mathrm{S}}\right) \\ &= \mathcal{B}_x s_y - \mathcal{B}_y s_x \end{aligned} \tag{4.21}$$

and

$$\begin{aligned} \dot{x}_{\mathrm{S}} &= -\frac{p_{\mathrm{S}}}{\sqrt{1-p_{\mathrm{S}}^2}} (\mathcal{B}_x \cos x_{\mathrm{S}} + \mathcal{B}_y \sin x_{\mathrm{S}}) + \mathcal{B}_z \\ &= \frac{s_z}{s_z^2 - 1} (\mathcal{B}_x s_x + \mathcal{B}_y s_y) + \mathcal{B}_z \, . \end{aligned} \tag{4.22}$$

Using $|s| = 1$ we also conclude that

$$\begin{aligned} \dot{s}_x &= \frac{-p_{\mathrm{S}}}{\sqrt{1-p_{\mathrm{S}}^2}} \, \dot{p}_{\mathrm{S}} \cos x_{\mathrm{S}} - \sqrt{1-p_{\mathrm{S}}^2}\, \dot{x}_{\mathrm{S}} \sin x_{\mathrm{S}} \\ &= \frac{s_z}{1-s_z^2} \mathcal{B}_y (s_x^2 + s_y^2) - \mathcal{B}_z s_y \\ &= \mathcal{B}_y s_z - \mathcal{B}_z s_y \end{aligned} \tag{4.23}$$

and analogously $\dot{s}_y = \mathcal{B}_z s_x - \mathcal{B}_x s_z$. Therefore, we find that classical spin precession,

$$\dot{\boldsymbol{s}} = \boldsymbol{\mathcal{B}} \times \boldsymbol{s} \,, \tag{4.24}$$

can be interpreted as Hamiltonian dynamics on S^2 with the Hamiltonian (4.19). However, the spin dynamics is not autonomous but driven by the translational dynamics and the skew product flow Y_{cl}^t, in contrast to the translational dynamics ϕ^t, in general is not Hamiltonian. The last statement can be seen as follows: If one wants to construct a joint Hamiltonian $\tilde{H}(\boldsymbol{p}, \boldsymbol{x}, \boldsymbol{s})$ for both spin and translational degrees of freedom it must have the form

$$\tilde{H}(\boldsymbol{p}, \boldsymbol{x}, \boldsymbol{s}) = \boldsymbol{s}\boldsymbol{\mathcal{B}}(\boldsymbol{p}, \boldsymbol{x}) + H_{\mathrm{trans}}(\boldsymbol{p}, \boldsymbol{x}) \tag{4.25}$$

in order to generate the spin precession, cf. (4.19). However, if $\boldsymbol{\mathcal{B}}$ is not constant the equations of motion for $\boldsymbol{p}$ and $\boldsymbol{x}$ will always involve a term proportional to $\boldsymbol{s}$ which contradicts our definition of Y_{cl}^t.

4.2 Excursion: Observables for Spinning Particles

We are now aiming at characterising the degree of chaoticity or regularity of the skew product Y_{cl}^t. Instead of giving the definition of properties like ergodicity or mixing in the usual way (e.g., defining ergodicity of Y_{cl}^t with respect to $\mu_{\mathrm{cl}} = \mu_E \times \mu_{S^2}$ by all invariant measurable subsets $A \subseteq \Omega_E \times S^2$, i.e. $Y_{\mathrm{cl}}^t(A) = A$, having either zero or full measure) we prefer to use equivalent formulations involving observables $a(\boldsymbol{p}, \boldsymbol{x}, \boldsymbol{s})$, which perhaps have a more immediate physical interpretation. Ergodicity of Y_{cl}^t with respect to μ_{cl} then means that for every observable $a(\boldsymbol{p}, \boldsymbol{x}, \boldsymbol{s})$ time averages equal phase space averages, i.e.

$$\lim_{T\to\infty} \frac{1}{T} \int_0^T a(Y_{\mathrm{cl}}^t(\boldsymbol{p}, \boldsymbol{x}, \boldsymbol{s}))\,\mathrm{d}t = \int_{\Omega_E \times S^2} a(\boldsymbol{p}', \boldsymbol{y}', \boldsymbol{s}')\,\mathrm{d}\mu_{\mathrm{cl}}(\boldsymbol{p}', \boldsymbol{y}', \boldsymbol{s}'))\,, \tag{4.26}$$

for μ_{cl}-almost all initial conditions $(\boldsymbol{p}, \boldsymbol{x}, \boldsymbol{s}) \in \Omega_E \times S^2$. In order to really give (4.26) a meaning one should specify from which space the observable may be chosen. Usually one would allow all integrable functions, i.e. $a \in L^1(\mathbb{R}^d \times \mathbb{R}^d \times S^2)$, see e.g. [1, 3].

However, if we think of classical observables as some limit of quantum observables (e.g., we have in a certain sense associated the classical Hamiltonian $\boldsymbol{s}\boldsymbol{\mathcal{B}}$, see (4.19), with the spin part $\hat{\boldsymbol{s}}\hat{\boldsymbol{\mathcal{B}}}$ of the Pauli Hamiltonian) the following problem arises. A quantum observable depending only on the spin degrees of freedom is a hermitian $(2s+1) \times (2s+1)$ matrix, an element of a vector space of (real) dimension $(2s+1)^2$. On the other hand the space $L^1(S^2)$ is infinite dimensional.

This contradiction can be illustrated as follows. Consider the case of a spin $1/2$. We would like to associate with the quantum operator $\hat{\boldsymbol{s}} = (\hbar/2)\boldsymbol{\sigma}$ the "classical" vector $(\hbar/2)\boldsymbol{s}$, $|\boldsymbol{s}| = 1$, and with the unit matrix $\mathbb{1}_2$ the constant function 1 on S^2. However, in quantum mechanics we have the identity $\hat{s}_x^2 = (\hbar^2/4)\mathbb{1}_2$ which we somehow have to implement for our classical observables on S^2 where s_x^2 is linearly independent of 1. One could be tempted to identify even powers of s_j with multiples of 1 and odd powers with multiples of s_j itself. Unfortunately, the situation changes if we want to consider a spin $s > 1/2$. Then $\hat{s}_j^2$ is not a multiple of $\mathbb{1}_{2s+1}$. For instance for $s = 1$ we have, cf. (3.10),

$$\hat{s}_x^2 = \left[\frac{\hbar}{2}\sqrt{2}\begin{pmatrix} 0 & 1 & 0 \\ 1 & 0 & 1 \\ 0 & 1 & 0 \end{pmatrix}\right]^2 = \frac{\hbar^2}{2}\begin{pmatrix} 1 & 0 & 1 \\ 0 & 2 & 0 \\ 1 & 0 & 1 \end{pmatrix} , \tag{4.27}$$

which is linearly independent of $\{\mathbb{1}_3, \hat{s}_x, \hat{s}_y, \hat{s}_z\}$. So we should not identify s_x^2 with a multiple of 1.

The solution to this dilemma is as follows. A classical observable, represented by a function on S^2, which corresponds to a quantum observable for a spin s may only be taken from a finite dimensional subspace of $L^1(S^2)$. For example, for a spin 1/2 this space is spanned by $\{1, s_x, s_y, s_z\}$. Then it can be shown that there exists a correspondence between classical and quantum observables [4, 5] with similar properties as the usual Wigner-Weyl correspondence for the translational degrees of freedom, cf. Appendix E. Since we will not use any particular properties of this Wigner-Weyl correspondence for spinning particles in the present work, we refrain from listing any details but refer the reader to the articles [4, 5] and also [6] where the formalism is employed in a semiclassical context and also some of its properties are reviewed. For our purposes it is sufficient to know that classical observables $a(\boldsymbol{s})$ for a spin s have an expansion in spherical harmonics $Y_{lm}(\boldsymbol{s})$ that terminates at $l = 2s$, i.e.

$$a(\boldsymbol{s}) = \sum_{l=0}^{2s} \sum_{m=-l}^{l} c_{lm} \, Y_{lm}(\boldsymbol{s}) , \tag{4.28}$$

where the coefficients c_{lm} are restricted to those values which lead to real valued observables $a(\boldsymbol{s})$. We denote this space of possible classical observables for a spin s by $\mathcal{H}_s \subset L^1(S^2)$. We will not base any results on this insight but only use it in Sect. 6.4 as a motivation for considering classical time evolution operators on $\mathcal{H}_s$ rather than on $L^2(S^2)$, which we are free to do anyway. Let us remark that the dimension of $\mathcal{H}_s$ is given by

$$\dim \mathcal{H}_s = \sum_{l=0}^{2s} \sum_{m=l}^{l} 1 = \sum_{l=0}^{2s} (2l+1) = (2s+1)^2 \tag{4.29}$$

which is consistent with the (real) dimension of the space of hermitian $(2s+1) \times (2s+1)$ matrices.

4.3 Ergodic Properties of the Skew Product

We briefly list what ergodicity and the mixing property of the skew product imply for the translational and spin dynamics, respectively.

If the skew product flow Y_{cl}^t is ergodic, then for all observables $a \in L^1(\mathbb{R}^d \times \mathbb{R}^d) \otimes \mathcal{H}_s$ and μ_{cl}-almost all $(\boldsymbol{p}, \boldsymbol{x}, \boldsymbol{s}) \in \Omega_E \times S^2$ the relation (4.26) holds. If the observable a does not depend on the spin degrees of freedom this equation simplifies to

$$\lim_{T\to\infty} \frac{1}{T} \int_0^T a(\phi^t(\boldsymbol{p}, \boldsymbol{x}))\, \mathrm{d}t = \int_{\Omega_E} a(\boldsymbol{p}', \boldsymbol{y}')\, \mathrm{d}\mu(\boldsymbol{p}', \boldsymbol{y}') \,, \tag{4.30}$$

for μ_E-almost all $(\boldsymbol{p}, \boldsymbol{x}) \in \Omega_E$. This means that the translational dynamics ϕ^t itself is ergodic, which is a consequence of Y_{cl}^t being a skew product with base flow ϕ^t. If, on the other hand, a is a function of $\boldsymbol{s}$ only, we have

$$\lim_{T\to\infty} \frac{1}{T} \int_0^T a(R(\boldsymbol{p}, \boldsymbol{x}, t)\boldsymbol{s})\, \mathrm{d}t = \int_{S^2} a(\boldsymbol{s}')\, \mathrm{d}\mu_{S^2}(\boldsymbol{s}') \tag{4.31}$$

for μ-almost all $(\boldsymbol{p}, \boldsymbol{x}, \boldsymbol{s}) \in \Omega_E \times S^2$, which corresponds to an equidistribution of $\boldsymbol{s}(t) = R(\boldsymbol{p}, \boldsymbol{x}, t)\boldsymbol{s}$ on S^2.

A system is called (strongly) mixing if correlations decay for all observables a_1, a_2, i.e.

$$\begin{aligned} \lim_{t\to\infty} \int_{\Omega_E \times S^2} & a_1(Y^t(\boldsymbol{p}, \boldsymbol{x}, \boldsymbol{s}))\, a_2(\boldsymbol{p}, \boldsymbol{x}, \boldsymbol{s})\, \mathrm{d}\mu_{\mathrm{cl}}(\boldsymbol{p}, \boldsymbol{x}, \boldsymbol{s}) \\ &= \int_{\Omega_E \times S^2} a_1(\boldsymbol{p}, \boldsymbol{x}, \boldsymbol{s})\, \mathrm{d}\mu_{\mathrm{cl}}(\boldsymbol{p}, \boldsymbol{x}, \boldsymbol{s}) \int_{\Omega_E \times S^2} a_2(\boldsymbol{p}', \boldsymbol{x}', \boldsymbol{s}')\, \mathrm{d}\mu_{\mathrm{cl}}(\boldsymbol{p}', \boldsymbol{x}', \boldsymbol{s}') \,. \end{aligned} \tag{4.32}$$

Again for observables a_1, a_2 which are independent of the spin degrees of freedom this implies the mixing property of ϕ^t,

$$\begin{aligned} \lim_{t\to\infty} \int_{\Omega_E} & a_1(\phi^t(\boldsymbol{p}, \boldsymbol{x}))\, a_2(\boldsymbol{p}, \boldsymbol{x})\, \mathrm{d}\mu_E(\boldsymbol{p}, \boldsymbol{x}) \\ &= \int_{\Omega_E} a_1(\boldsymbol{p}, \boldsymbol{x})\, \mathrm{d}\mu_E(\boldsymbol{p}, \boldsymbol{x}) \int_{\Omega_E} a_2(\boldsymbol{p}', \boldsymbol{x}')\, \mathrm{d}\mu_E(\boldsymbol{p}', \boldsymbol{x}') \,. \end{aligned} \tag{4.33}$$

since Y^t is a skew product. On the other hand, for a pure spin observable we find

$$\begin{aligned} \lim_{t\to\infty} \int_{\Omega_E \times S^2} & a_1(R(\boldsymbol{p}, \boldsymbol{x}, t)\boldsymbol{s})\, a_2(\boldsymbol{s})\, \mathrm{d}\mu_{\mathrm{cl}}(\boldsymbol{p}, \boldsymbol{x}, \boldsymbol{s}) \\ &= \int_{S^2} a_1(\boldsymbol{s})\, \mathrm{d}\mu_{S^2}(\boldsymbol{s}) \int_{S^2} a_2(\boldsymbol{s}')\, \mathrm{d}\mu_{S^2}(\boldsymbol{s}') \,, \end{aligned} \tag{4.34}$$

i.e. the spin vectors $\boldsymbol{s}(t) = R(\boldsymbol{p}, \boldsymbol{x}, t)\boldsymbol{s}$ and $\boldsymbol{s}(0) = \boldsymbol{s}$ become uncorrelated for large times t.

4.4 Integrable Systems

We review the notion of integrability for Hamiltonian systems and formulate the main theorem due to Liouville and Arnold which states under which conditions a Hamiltonian system is integrable. In Sect. 4.4.2 we extend the notion of integrability to the skew product of translational and spin degrees of freedom and prove the respective generalisation of the Theorem of Liouville and Arnold.

4.4.1 Hamiltonian Systems – the Theorem of Liouville and Arnold

A system of ordinary differential equations is called integrable if one can find sufficiently many (independent) first integrals, i.e. functions which are invariant under the time evolution of the system. For a system of $2d$ first order differential equations one needs to find $2d$ integrals in terms of which the solution of the system can be expressed. A Hamiltonian system for a mechanical problem with d degrees of freedom has a $2d$-dimensional phase space on which the dynamics is governed by a system of $2d$ first oder differential equations, Hamilton's equations of motion. For this situation Liouville showed [7] that it is sufficient to know d such integrals of motion, if in addition all their Poisson brackets vanish pairwise. Then the system can be solved by quadratures, i.e. by calculating integrals and inverting known functions.

A modern formulation of this theorem together with a complete proof is due to Arnold [8]. He showed that under fairly general conditions the motion of such a system is conditionally periodic and also stressed the underlying geometry: The motion of an integrable system takes place on d-dimensional tori in phase space. Let us cite Arnold's version of the theorem as stated in [8, Chap. 10].

Theorem 1. (Liouville, Arnold) *Suppose that we are given d (smooth) functions in involution on a symplectic $2d$-dimensional manifold,*

$$A_1 = H, A_2, \dots, A_d \quad \textit{with} \quad \{A_j, A_k\} = 0 \ , \ j,k = 1,2,\dots,d \ . \tag{4.35}$$

Consider the level set of the functions A_j,

$$M_{\boldsymbol{a}} = \{(\boldsymbol{p}, \boldsymbol{x}) \,|\, A_j(\boldsymbol{p}, \boldsymbol{x}) = a_j, \ j = 1, \dots, d\} \ . \tag{4.36}$$

Assume that the d functions A_j are independent on $M_{\boldsymbol{a}}$ (i.e., the d 1-forms $\mathrm{d}A_j$ are linearly independent at each point of $M_{\boldsymbol{a}}$). Then

1. *$M_{\boldsymbol{a}}$ is a smooth manifold, invariant under the phase flow with Hamiltonian $H = A_1$.*
2. *If the manifold $M_{\boldsymbol{a}}$ is compact and connected, then it is diffeomorphic to the d-dimensional torus*

$$\mathbb{T}^d = \{(\vartheta_1, \dots, \vartheta_d) \mod 2\pi\} \ . \tag{4.37}$$

3. *The phase flow with Hamiltonian H determines a conditionally periodic motion on $M_{\boldsymbol{a}}$, i.e., in angular coordinates $\boldsymbol{\vartheta} = (\vartheta_1, \ldots, \vartheta_d)$ we have*

$$\frac{\mathrm{d}\boldsymbol{\vartheta}}{\mathrm{d}t} = \boldsymbol{\omega} \, , \quad \boldsymbol{\omega} = \boldsymbol{\omega}(\boldsymbol{a}) \, . \tag{4.38}$$

4. *The canonical equations with the Hamiltonian H can be integrated by quadratures.*

We remark that we define the Poisson bracket of two observables $A(\boldsymbol{p}, \boldsymbol{x})$, $B(\boldsymbol{p}, \boldsymbol{x})$ as

$$\{A, B\} = \sum_{j=1}^{d} \left(\frac{\partial A}{\partial p_j} \frac{\partial B}{\partial x_j} - \frac{\partial A}{\partial x_j} \frac{\partial B}{\partial p_j} \right) \, . \tag{4.39}$$

We refrain from giving a complete proof of this theorem but instead refer the reader to Arnold's book [8] where the theorem is proven in great detail. The key step is to show that given the conditions (4.35) the d flows ϕ_j^t with Hamiltonians A_j, $j = 1, \ldots, d$, commute, i.e.

$$\phi_j^t \circ \phi_k^{t'} = \phi_k^{t'} \circ \phi_j^t \quad \forall \, t, t' \in \mathbb{R} \quad \text{and} \quad j, k = 1, \ldots, d \, . \tag{4.40}$$

Starting from this insight Arnold constructs the properties 1-3 as listed in Theorem 1.

In oder to prove property 4 we remark that under the conditions of Theorem 1 the system can be transformed to action and angle variables. To this end notice that using the coordinates $(\boldsymbol{A}, \boldsymbol{\vartheta})$ on phase space the equations of motion are simply given by

$$\frac{\mathrm{d}\boldsymbol{A}}{\mathrm{d}t} = 0 \, , \quad \frac{\mathrm{d}\boldsymbol{\vartheta}}{\mathrm{d}t} = \boldsymbol{\omega}(\boldsymbol{A}) \, . \tag{4.41}$$

However, in general $\boldsymbol{A}$ and $\boldsymbol{\vartheta}$ are not canonically conjugate. Instead, one can find new constants of motion $\boldsymbol{I} = \boldsymbol{I}(\boldsymbol{A})$ such that $(\boldsymbol{I}, \boldsymbol{\vartheta})$ are a set of canonically conjugate coordinates $\boldsymbol{\vartheta}$ and momenta $\boldsymbol{I}$. The latter are called action variables, whereas the $\boldsymbol{\vartheta}$ are referred to as angle variables. If we express the Hamiltonian $H(\boldsymbol{p}, \boldsymbol{x})$ in terms of the new coordinates it becomes a function $\overline{H}(\boldsymbol{I})$ which does not depend on the coordinates $\boldsymbol{\vartheta}$ (since the actions $\boldsymbol{I}$ are conserved), and Hamilton's equations of motion read

$$\frac{\mathrm{d}\boldsymbol{\vartheta}}{\mathrm{d}t} = \nabla_{\boldsymbol{I}} \overline{H}(\boldsymbol{I}) = \boldsymbol{\omega} \, , \quad \frac{\mathrm{d}\boldsymbol{I}}{\mathrm{d}t} = -\nabla_{\boldsymbol{\vartheta}} \overline{H}(\boldsymbol{I}) = 0 \, . \tag{4.42}$$

The action variables $\boldsymbol{I}$ can be found as follows. On a torus $\mathbb{T}^d$ choose a basis of cycles $C_1, \ldots, C_d$ such that along C_j the angles ϑ_k, $k \neq j$, are constant and ϑ_j changes by 2π. Then

$$I_j = \frac{1}{2\pi} \oint_{C_j} \boldsymbol{p} \, \mathrm{d}\boldsymbol{x} \tag{4.43}$$

is an integral of motion canonically conjugate to ϑ_j. All steps which need to be performed in order to find the action and angle variables $(\boldsymbol{I}, \boldsymbol{\vartheta})$ are quadratures. Again we do not give any details but refer the reader to [8] or any other textbook on classical mechanics, like [9].

4.4.2 Integrability of the Skew Product

We now generalise the results for integrable Hamiltonian flows, which were reviewed in the preceeding section, to the case of the skew product Y_{cl}^t : $\mathbb{R}^d \times \mathbb{R}^d \times S^2 \to \mathbb{R}^d \times \mathbb{R}^d \times S^2$, which arises in the semiclassical treatment of the Pauli and the Dirac equation, see [10, 11]. To this end we first need a generalisation of the property (4.35) of two observables $A_1(\boldsymbol{p}, \boldsymbol{x})$ and $A_2(\boldsymbol{p}, \boldsymbol{x})$ being in involution, i.e. $\{A_1, A_2\} = 0$. A good generalisation of this condition should give rise to commuting skew products of the form of Y_{cl}^t, in a similar way as condition (4.35) gave rise to commuting Hamiltonian flows, cf. (4.40). In a second step we will determine simultaneous invariant manifolds – submanifolds of $\mathbb{R}^d \times \mathbb{R}^d \times S^2$ – of a given set of commuting skew products. We will see that in order to obtain a similar result as in Sect. 4.4.1 it is still sufficient to have d commuting skew products, although the symplectic manifold on which the system evolves, i.e. the extended phase space $\mathbb{R}^d \times \mathbb{R}^d \times S^2$, now has dimension $2d + 2$ rather than $2d$ as in the preceeding section. The invariant manifolds will turn out to be bundles over Liouville–Arnold tori $\mathbb{T}^d$ with typical fibre S^1. Moreover the fibre attached at a given point of the torus will be characterised by a latitude on the sphere S^2. Let us start with the following definition.

Definition 1. *The skew product flow Y_{cl}^t is called integrable if*

(i) the Hamiltonian flow ϕ_H^t is integrable, i.e. besides the Hamiltonian $H(\boldsymbol{p}, \boldsymbol{x}) =: A_1(\boldsymbol{p}, \boldsymbol{x})$ there are $d-1$ more independent integrals of motion, $A_2(\boldsymbol{p}, \boldsymbol{x}), \ldots, A_d(\boldsymbol{p}, \boldsymbol{x})$, with

$$\{A_j, A_k\} = 0 , \tag{4.44}$$

and

(ii) the corresponding flows ϕ_j^t can be extended to skew products ${Y_{\text{cl}}}_j^t$ on $\mathbb{R}^d \times \mathbb{R}^d \times S^2$ with fields $\boldsymbol{\mathcal{B}}_j$, i.e.

$${Y_{\text{cl}}}_j^t(\boldsymbol{p}, \boldsymbol{x}, \boldsymbol{s}(0)) = (\phi_j^t(\boldsymbol{p}, \boldsymbol{x}), \boldsymbol{s}(t)) , \quad \dot{\boldsymbol{s}}(t) = \boldsymbol{\mathcal{B}}_j(\phi_j^t(\boldsymbol{p}, \boldsymbol{x})) \times \boldsymbol{s}(t) , \tag{4.45}$$

for which we have

$$\{A_j, \boldsymbol{\mathcal{B}}_k\} + \{\boldsymbol{\mathcal{B}}_j, A_k\} - \boldsymbol{\mathcal{B}}_j \times \boldsymbol{\mathcal{B}}_k = 0 . \tag{4.46}$$

Let us first emphasize by the following lemma that this is a good definition.

Lemma 1. *Two skew products $Y_{\mathrm{cl}}{}_j^t$, $Y_{\mathrm{cl}}{}_k^t$ of the type (4.45) commute, if and only if the corresponding base flows commute and if*

$$\{A_j, \mathcal{B}_k\} + \{\mathcal{B}_j, A_k\} - \mathcal{B}_j \times \mathcal{B}_k = 0 \,. \tag{4.47}$$

Proof. Obviously the skew products can only commute if the corresponding base flows commute. Thus it remains to show that

$$R_j(\phi_k^t(\boldsymbol{p},\boldsymbol{x}),t')\, R_k(\boldsymbol{p},\boldsymbol{x},t)\boldsymbol{s} = R_k(\phi_j^{t'}(\boldsymbol{p},\boldsymbol{x}),t)\, R_j(\boldsymbol{p},\boldsymbol{x},t')\boldsymbol{s} \tag{4.48}$$

for all $(\boldsymbol{p},\boldsymbol{x},\boldsymbol{s}) \in \mathbb{R}^d \times \mathbb{R}^d \times S^2$ and $t,t' \in \mathbb{R}$ if and only if (4.47) holds. Here $R_j(\boldsymbol{p},\boldsymbol{x},t)\boldsymbol{s}(0)$ is the solution of

$$\dot{\boldsymbol{s}}(t) = \mathcal{B}_j(\phi_j^t(\boldsymbol{p},\boldsymbol{x})) \times \boldsymbol{s}(t) \,. \tag{4.49}$$

Certainly the condition (4.48) is equivalent to the vanishing of the difference

$$R_j(\phi_k^t(\boldsymbol{p},\boldsymbol{x}),t')\, R_k(\boldsymbol{p},\boldsymbol{x},t) - R_k(\phi_j^{t'}(\boldsymbol{p},\boldsymbol{x}),t)\, R_j(\boldsymbol{p},\boldsymbol{x},t') \,. \tag{4.50}$$

Moreover, since $R_j \in \mathrm{SO}(3)$ is obtained from $d_j \in \mathrm{SU}(2)$ with

$$\dot{d}_j(\boldsymbol{p},\boldsymbol{x},t) + \frac{\mathrm{i}}{2}\boldsymbol{\sigma}\mathcal{B}_j(\phi_j^t(\boldsymbol{p},\boldsymbol{x}))\, d_j(\boldsymbol{p},\boldsymbol{x},t) = 0 \,, \quad d_j(\boldsymbol{p},\boldsymbol{x},0) = \mathbb{1}_2 \tag{4.51}$$

by means of the universal covering map, cf. Appendix C, we may instead investigate the difference[1]

$$\Delta(t,t') := d_j(\phi_k^t(\boldsymbol{p},\boldsymbol{x}),t')\, d_k(\boldsymbol{p},\boldsymbol{x},t) - d_k(\phi_j^{t'}(\boldsymbol{p},\boldsymbol{x}),t)\, d_j(\boldsymbol{p},\boldsymbol{x},t') \,. \tag{4.52}$$

One immediately finds that

$$\Delta(t,0) = 0 = \Delta(0,t') \quad \forall\, t,t' \in \mathbb{R} \,, \tag{4.53}$$

and thus

$$\Delta(t,t') = \frac{\partial^2 \Delta}{\partial t \partial t'}(0,0)\, tt' + \mathrm{o}\left(t^2 + t'^2\right) \,, \quad t,t' \to 0 \,. \tag{4.54}$$

We therefore need to calculate this mixed second derivative,

[1] One could argue that since φ is a double covering, $\varphi(-g) = \varphi(g)$, we could have vanishing (4.50) without vanishing $\Delta(t,t')$; however, since our proof below constructs all properties continuously starting from $t = t' = 0$ this case can be excluded.

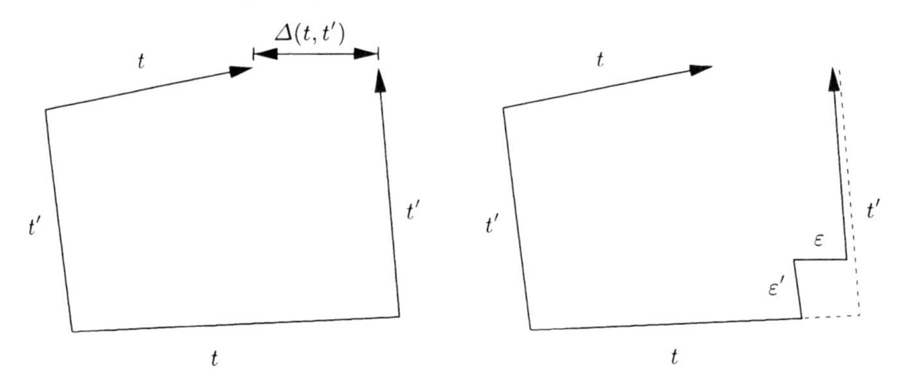

Fig. 4.1. Schematic illustration of the difference $\Delta(t,t')$ defined in (4.52) and of the path deformations described below (4.56).

$$
\begin{aligned}
\frac{\partial^2 \Delta}{\partial t \partial t'}(0,0) &= \frac{\partial}{\partial t}\left[-\frac{\mathrm{i}}{2}\boldsymbol{\sigma}\boldsymbol{\mathcal{B}}_j(\phi_k^t(\boldsymbol{p},\boldsymbol{x}))\, d_k(\boldsymbol{p},\boldsymbol{x},t)\right]_{t=0} \\
&\quad - \frac{\partial}{\partial t'}\left[-\frac{\mathrm{i}}{2}\boldsymbol{\sigma}\boldsymbol{\mathcal{B}}_k(\phi_j^{t'}(\boldsymbol{p},\boldsymbol{x}))\, d_j(\boldsymbol{p},\boldsymbol{x},t')\right]_{t'=0} \\
&= \left[-\frac{\mathrm{i}}{2}\boldsymbol{\sigma}\{A_k,\boldsymbol{\mathcal{B}}_j\}(\boldsymbol{p},\boldsymbol{x}) + \frac{\mathrm{i}}{2}\left(\boldsymbol{\sigma}\boldsymbol{\mathcal{B}}_j(\boldsymbol{p},\boldsymbol{x})\right)\frac{\mathrm{i}}{2}\left(\boldsymbol{\sigma}\boldsymbol{\mathcal{B}}_k(\boldsymbol{p},\boldsymbol{x})\right)\right] \\
&\quad - \left[-\frac{\mathrm{i}}{2}\boldsymbol{\sigma}\{A_j,\boldsymbol{\mathcal{B}}_k\}(\boldsymbol{p},\boldsymbol{x}) + \frac{\mathrm{i}}{2}\left(\boldsymbol{\sigma}\boldsymbol{\mathcal{B}}_k(\boldsymbol{p},\boldsymbol{x})\right)\frac{\mathrm{i}}{2}\left(\boldsymbol{\sigma}\boldsymbol{\mathcal{B}}_j(\boldsymbol{p},\boldsymbol{x})\right)\right] \\
&= \frac{\mathrm{i}}{2}\boldsymbol{\sigma}\left[\{\boldsymbol{\mathcal{B}}_j, A_k\} + \{A_j,\boldsymbol{\mathcal{B}}_k\} - \boldsymbol{\mathcal{B}}_j \times \boldsymbol{\mathcal{B}}_k\right](\boldsymbol{p},\boldsymbol{x}) \, .
\end{aligned}
\tag{4.55}
$$

Hence, we see that commuting skew products imply (4.47) which proves the first half of Lemma 1. We still have to show the reverse direction in which case we now have

$$
\Delta(t,t') = \mathrm{o}\left(t^2 + t'^2\right) , \quad t, t' \to 0 \, . \tag{4.56}
$$

That this property indeed implies that the flows commute can be seen as follows: In Fig. 4.1 we have schematically visualised the difference $\Delta(t,t')$ between first following $Y_{\mathrm{cl}_j}^{\,t}$ and then $Y_{\mathrm{cl}_k}^{\,t'}$, and the other way round. Breaking up the time intervals into N parts of length $\varepsilon = t/N$, $\varepsilon' = t'/N$, we can deform the two paths into each other by N^2 small changes, the first of which is illustrated in Fig. 4.1. Each of these steps introduces an error of the order $\mathrm{o}\left(\varepsilon^2 + \varepsilon'^2\right) = \mathrm{o}\left(1/N^2\right)$ and, since we need N^2 steps, we find

$$
\Delta(t,t') = N^2\, \mathrm{o}\left(\frac{1}{N^2}\right) = \mathrm{o}\,(1) \, , \quad N \to \infty \, , \tag{4.57}
$$

i.e. $\Delta(t,t')$ vanishes, which proves Lemma 1.

We now state the main result of this section in the following theorem.

Theorem 2. *If the skew product flow Y_{cl}^t is integrable in the sense of Definition 1, then the combined phase space can be decomposed into invariant bundles $\mathcal{T}_\theta \xrightarrow{\pi} \mathbb{T}^d$ over Liouville–Arnold tori with typical fibre S^1. The bundles can be embedded in $\mathbb{T}^d \times S^2$ such that the fibres are characterised by their latitude θ with respect to a local direction $\boldsymbol{n}(\boldsymbol{p}, \boldsymbol{x}) \in S^2$, i.e.*

$$\mathcal{T}_\theta = \left\{(\boldsymbol{p}, \boldsymbol{x}, \boldsymbol{s}) \in \mathbb{T}^d \times S^2 \mid \sphericalangle(\boldsymbol{s}, \boldsymbol{n}(\boldsymbol{p}, \boldsymbol{x})) = \theta\right\} . \tag{4.58}$$

Instead of directly proving Theorem 2 we prefer to work out the corresponding result for the flow Y^t on $\mathbb{R}^d \times \mathbb{R}^d \times \mathrm{SU}(2)$ first. In this case the invariant manifolds are obtained by attaching a one-parameter subgroup of SU(2) to each point of the torus. By means of the covering map $\varphi : \mathrm{SU}(2) \to \mathrm{SO}(3)$ this yields a one-parameter subgroup of rotations which can be characterised by a direction $\boldsymbol{n}$. Applying this subgroup to a given spin vector $\boldsymbol{s}$ we obtain the following subset of S^2 which is characterised by the latitude with respect to $\boldsymbol{n}$,

$$\{\boldsymbol{s} \in S^2 \mid \sphericalangle(\boldsymbol{s}, \boldsymbol{n}) = const\} . \tag{4.59}$$

Proof. As we have seen previously, integrability of Y_{cl}^t also implies integrability of Y^t in the sense that the conditions (i) and (ii) of Definition 1 ensure not only that the Hamiltonian flows ϕ_j^t are integrable but also that the corresponding SU(2)-extensions commute. We may therefore define a multi-time flow by

$$\mathbb{Y}^{\boldsymbol{t}} := Y_d^{t_d} \circ \cdots \circ Y_2^{t_2} \circ Y_1^{t_1} , \tag{4.60}$$

where due to commutativity the ordering is not relevant. Explicitly $\mathbb{Y}^{\boldsymbol{t}}$ maps a point $(\boldsymbol{p}, \boldsymbol{x}, g) \in \mathbb{R}^d \times \mathbb{R}^d \times \mathrm{SU}(2)$ to

$$\begin{aligned} &\mathbb{Y}^{\boldsymbol{t}}(\boldsymbol{p}, \boldsymbol{x}, g) = \left((\phi_d^{t_d} \circ \cdots \circ \phi_1^{t_1})(\boldsymbol{p}, \boldsymbol{x}),\, \mathbb{d}(\boldsymbol{p}, \boldsymbol{x}, \boldsymbol{t})g\right) \\ &\text{with} \quad \mathbb{d}(\boldsymbol{p}, \boldsymbol{x}, \boldsymbol{t}) := d_d\left((\phi_{d-1}^{t_{d-1}} \circ \cdots \circ \phi_1^{t_1})(\boldsymbol{p}, \boldsymbol{x}), t_d\right) \cdots d_1(\boldsymbol{p}, \boldsymbol{x}, t_1)g . \end{aligned} \tag{4.61}$$

Now restrict the translational dynamics to a Liouville–Arnold torus which is invariant under all ϕ_j^t, i.e. in the combined phase space we consider a submanifold $\mathbb{T}^d \times \mathrm{SU}(2)$. On $\mathbb{T}^d$ we can now chose d topologically different closed paths $\mathcal{C}_j$, such that topologically every closed loop on $\mathbb{T}^d$ is an integer linear combination of these basis cycles. Since the base flow ϕ_1^t is integrable we can associate with each of the basis cycles $\mathcal{C}_j$ a unique tuple $\boldsymbol{t}_j$ such that the projection of a flow line along $\mathbb{Y}^{\boldsymbol{t}_j}$ is the cycle $\mathcal{C}_j$. The corresponding cocycle we denote by

$$d_{\mathcal{C}_j}(\boldsymbol{p}, \boldsymbol{x}) := \mathbb{d}(\boldsymbol{p}, \boldsymbol{x}, \boldsymbol{t}_j) . \tag{4.62}$$

Now commutativity of the flows Y_j^t guarantees that all $d_{\mathcal{C}_j}(\boldsymbol{p}, \boldsymbol{x})$, $j = 1, \ldots, d$, at a given point $(\boldsymbol{p}, \boldsymbol{x}) \in \mathbb{T}^d$ commute. Since any two commuting elements

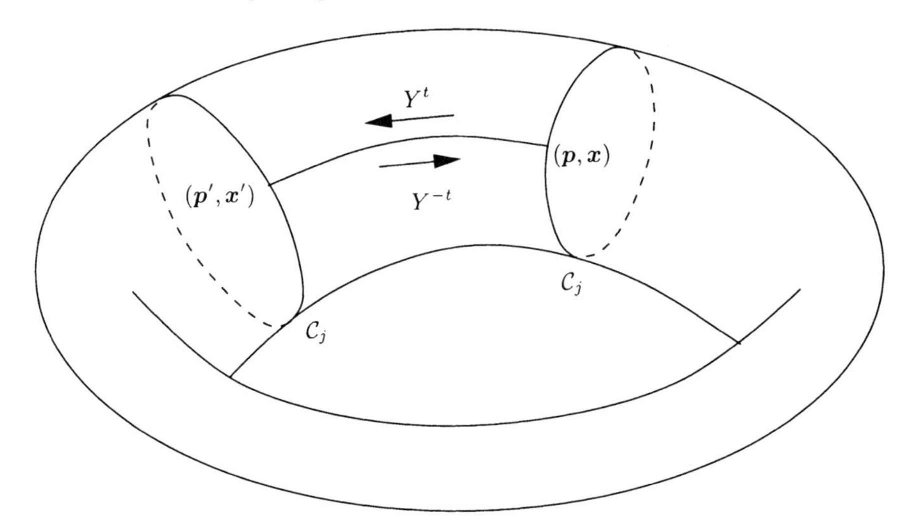

Fig. 4.2. Two topologically identical representations of the basis cycle $\mathcal{C}_j$ whose starting points $(\boldsymbol{p}, \boldsymbol{x})$ and $(\boldsymbol{p}', \boldsymbol{x}')$, respectively, are connected by a flow line of Y^t.

of SU(2) have to lie in the same one-parameter subgroup, we can associate with the point $(\boldsymbol{p}, \boldsymbol{x}) \in \mathbb{T}^d$ the subgroup

$$G_{\boldsymbol{n}(\boldsymbol{p},\boldsymbol{x})} := \left\{ g \in \mathrm{SU}(2) \,|\, g = \mathrm{e}^{-\mathrm{i}\frac{\alpha}{2}\boldsymbol{\sigma n}(\boldsymbol{p},\boldsymbol{x})} \,,\; \alpha \in [0, 4\pi) \right\} \tag{4.63}$$

with some fixed direction $\boldsymbol{n}(\boldsymbol{p}, \boldsymbol{x})$, $|\boldsymbol{n}(\boldsymbol{p}, \boldsymbol{x})| = 1$, i.e

$$d_{\mathcal{C}_j}(\boldsymbol{p}, \boldsymbol{x}) \in G_{\boldsymbol{n}(\boldsymbol{p},\boldsymbol{x})} \quad \forall\, j = 1, \ldots, d\,. \tag{4.64}$$

Now consider two different points $(\boldsymbol{p}, \boldsymbol{x})$, $(\boldsymbol{p}', \boldsymbol{x}') \in \mathbb{T}^d$. Since we can reach any point on the torus by linear combinations of ϕ_j^t we have $(\boldsymbol{p}', \boldsymbol{x}') = (\pi \circ \mathbb{Y}^{\boldsymbol{t}})(\boldsymbol{p}, \boldsymbol{x})$ for some $\boldsymbol{t}$. From

$$\mathbb{Y}^{-\boldsymbol{t}} \circ \mathbb{Y}^{\boldsymbol{t}_j} \circ \mathbb{Y}^{\boldsymbol{t}} = \mathbb{Y}^{\boldsymbol{t}_j}\,, \tag{4.65}$$

cf. Fig. 4.2, we conclude that

$$d_{\mathcal{C}_j}(\boldsymbol{p}', \boldsymbol{x}') = g\, d_{\mathcal{C}_j}(\boldsymbol{p}, \boldsymbol{x})\, g^{-1} \quad \forall\, j = 1, \ldots, d \tag{4.66}$$

with a fixed $g = \mathbb{d}(\boldsymbol{p}, \boldsymbol{x}, \boldsymbol{t}) \in \mathrm{SU}(2)$. Thus, the one-parameter subgroups associated with two different points on the same torus are simply related by conjugation with a fixed element of SU(2). This corresponds to a rotation of the axis $\boldsymbol{n}$,

$$G_{\boldsymbol{n}(\boldsymbol{p}',\boldsymbol{x}')} = G_{\varphi(g)\,\boldsymbol{n}(\boldsymbol{p},\boldsymbol{x})}\,, \tag{4.67}$$

where $\varphi : \mathrm{SU}(2) \to \mathrm{SO}(3)$ is the covering map, cf. Appendix C. Since a one-parameter subgroup is isomorphic to U(1) we can also say that the invariant

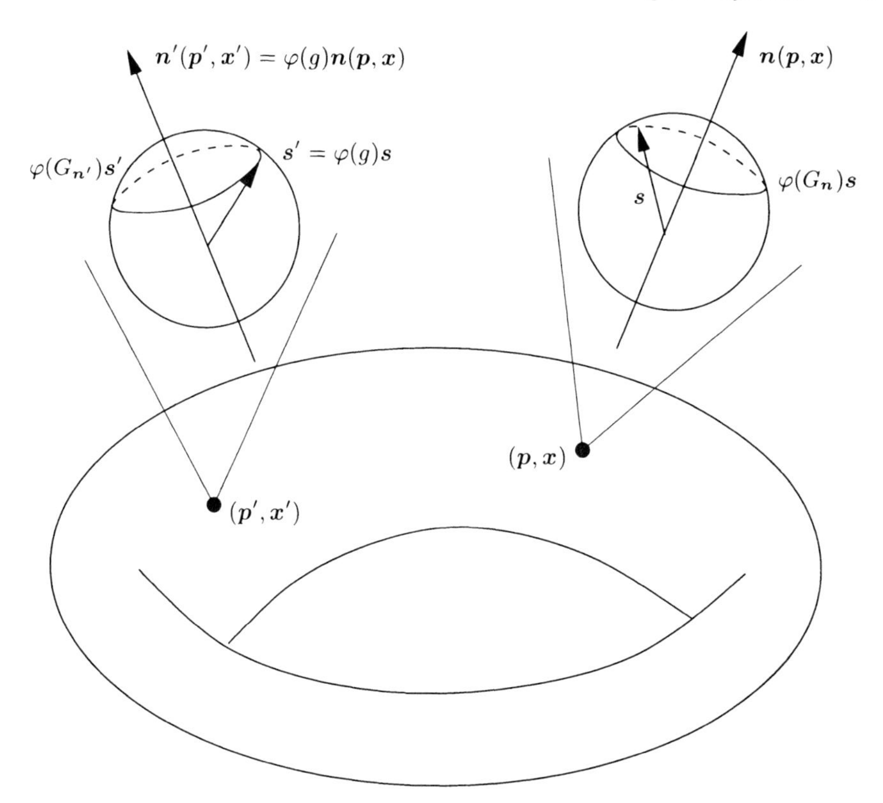

Fig. 4.3. The invariant manifolds $\mathcal{T}_\theta$, of $Y_{\rm cl}^t$, see (4.58), are given by tori $\mathbb{T}^d$ to which at each point is attached the set of all points on the two-sphere S^2 sharing a fixed latitude θ with respect to a varying axis $\boldsymbol{n}(\boldsymbol{p}, \boldsymbol{x})$.

manifolds of an integrable SU(2)-extension Y^t are fibre bundles over Liouville Arnold tori with $U(1)$-holonomy.

Notice that from (4.66) we also conclude that one can write

$$d_{\mathcal{C}_j}(\boldsymbol{p}, \boldsymbol{x}) = \exp\left(-\mathrm{i}\frac{\alpha_j}{2}\boldsymbol{\sigma}\boldsymbol{n}_j(\boldsymbol{p}, \boldsymbol{x})\right) , \tag{4.68}$$

since conjugation with $g \in \mathrm{SU}(2)$ only rotates $\boldsymbol{n}_j$ but leaves the angles α_j invariant. Thus the direction $\boldsymbol{n}_j$ is a function of the point $(\boldsymbol{p}, \boldsymbol{x}) \in \mathbb{T}^d$ but the angle α_j depends only on the topology of $\mathcal{C}_j$.

Having determined the invariant manifolds of Y^t we can now return to the purely classical skew product $Y_{\rm cl}^t$. For a point $(\boldsymbol{p}, \boldsymbol{x}, \boldsymbol{s}) \in \mathbb{T}^d \times S^2$ the invariant bundle which it lies on is constructed as follows (see Fig. 4.3 for an illustration). The fibre over $(\boldsymbol{p}, \boldsymbol{x})$ is obtained by applying $\varphi(G_{\boldsymbol{n}(\boldsymbol{p},\boldsymbol{x})})$ to $\boldsymbol{s}$. When now moving to a different point $(\boldsymbol{p}', \boldsymbol{x}') \in \mathbb{T}^d$ along the flow lines of ϕ_j^t the spin vector $\boldsymbol{s}$ is rotated by some $\varphi(g) \in \mathrm{SO}(3)$. The fibre over $(\boldsymbol{p}', \boldsymbol{x}')$ is then obtained by applying $\varphi(G_{\boldsymbol{n}(\boldsymbol{p}',\boldsymbol{x}')})$ to $\boldsymbol{s}' = \varphi(g)\boldsymbol{s}$. However, since $\boldsymbol{n}$ is rotated likewise, $\boldsymbol{n}(\boldsymbol{p}', \boldsymbol{x}') = \varphi(g)\, \boldsymbol{n}(\boldsymbol{p}, \boldsymbol{x})$, the angle

$$\theta := \sphericalangle(\boldsymbol{n}(\boldsymbol{p}, \boldsymbol{x}), \boldsymbol{s}) = \sphericalangle(\boldsymbol{n}(\boldsymbol{p}', \boldsymbol{x}'), \boldsymbol{s}') \tag{4.69}$$

is invariant. This concludes our proof of Theorem 2.

4.5 Reprise: Trace Formula for Integrable Systems

In Sect. 3.5.3 we have derived the following semiclassical trace formula for systems whose classical translational dynamics are integrable (3.118),

$$\begin{aligned}\sum_n \varrho\left(\frac{E_n - E}{\hbar}\right) &= \frac{\hat{\varrho}(0)}{2\pi} \frac{(2s+1)\,|\Omega_E|}{(2\pi\hbar)^{d-1}} [1 + \mathcal{O}(\hbar)] \\ &+ \sum_{\boldsymbol{m}\in\mathbb{Z}^d\setminus\{0\}} \frac{\hat{\varrho}(T_{\boldsymbol{m}})}{2\pi} \frac{(2\pi)^d \exp\left(-\mathrm{i}\frac{\pi}{2}\boldsymbol{m}\boldsymbol{\mu} - \mathrm{i}\frac{\pi}{4}(d-1)\mathrm{sign}T_{\boldsymbol{m}}\right)}{(2\pi\hbar)^{(d-1)/2}\,|\det \mathbb{D}_{\boldsymbol{m}}|^{1/2}} \overline{\chi_s}^{\boldsymbol{m}} \\ &\times \exp\left(\frac{\mathrm{i}}{\hbar} 2\pi \boldsymbol{m}\boldsymbol{I}\right) [1 + \mathcal{O}(\hbar)] \,.\end{aligned} \tag{4.70}$$

The average $\overline{\chi_s}^{\boldsymbol{m}}$ of the spin angles $\alpha(\boldsymbol{I}_{\boldsymbol{m}}, \boldsymbol{\vartheta}, T_{\boldsymbol{m}})$ for periodic trajectories starting at different points of a given rational torus, characterised by winding numbers $\boldsymbol{m}$, can now be calculated if not only ϕ_H^t but also Y_{cl}^t is integrable. Then, due to the commutativity of the flows $Y_{\mathrm{cl}\,1}^{\;t}, \ldots, Y_{\mathrm{cl}\,d}^{\;t}$, the angels $\alpha(\boldsymbol{I}_{\boldsymbol{m}}, \boldsymbol{\vartheta}, T_{\boldsymbol{m}})$ are identical for all $\boldsymbol{\vartheta}$, cf. 4.68, and it makes sense to define

$$\alpha_{\boldsymbol{m}} := \alpha(\boldsymbol{I}_{\boldsymbol{m}}, \boldsymbol{\vartheta}, T_{\boldsymbol{m}}) \,. \tag{4.71}$$

The angle $\alpha_{\boldsymbol{m}}$ is related to the angles α_j characterising spin transport along a basis cycle $\mathcal{C}_j$, which were defined in the preceeding section, by

$$\alpha_{\boldsymbol{m}} = \boldsymbol{\alpha}\,\boldsymbol{m} = \sum_{j=1}^{m} \alpha_j m_j \tag{4.72}$$

since a periodic orbit on a the torus with winding numbers $\boldsymbol{m}$ is topologically equivalent to following each cycle $\mathcal{C}_j$ exactly m_j times. Finally, if Y_{cl}^t is integrable, the trace formula reads

$$\begin{aligned}\sum_n \varrho\left(\frac{E_n - E}{\hbar}\right) &= \frac{\hat{\varrho}(0)}{2\pi} \frac{(2s+1)\,|\Omega_E|}{(2\pi\hbar)^{d-1}} [1 + \mathcal{O}(\hbar)] \\ &+ \sum_{\boldsymbol{m}\in\mathbb{Z}^d\setminus\{0\}} \frac{\hat{\varrho}(T_{\boldsymbol{m}})}{2\pi} \frac{(2\pi)^d \exp\left(-\mathrm{i}\frac{\pi}{2}\boldsymbol{m}\boldsymbol{\mu} - \mathrm{i}\frac{\pi}{4}(d-1)\mathrm{sign}T_{\boldsymbol{m}}\right)}{(2\pi\hbar)^{(d-1)/2}\,|\det \mathbb{D}_{\boldsymbol{m}}|^{1/2}} \chi_s(\alpha_{\boldsymbol{m}}) \\ &\times \exp\left(\frac{\mathrm{i}}{\hbar} 2\pi \boldsymbol{m}\boldsymbol{I}\right) [1 + \mathcal{O}(\hbar)] \,.\end{aligned} \tag{4.73}$$

References

1. I.P. Cornfeld, S.V. Fomin, Ya.G. Sinai: *Ergodic Theory* (Springer-Verlag, New York, 1982)
2. Ya.G. Sinai (Ed.): *Dynamical Systems II, Ergodic Theory with Applications to Dynamical Systems and Statistical Mechanics*, Vol. 2 of Encyclopaedia of Mathematical Sciences (Springer-Verlag, Berlin, 1989)
3. R. Mañé: *Ergodic Theory and Differentiable Dynamics* (Springer-Verlag, Berlin Heidelberg New York, 1987)
4. J.M. Gracia-Bondía, J.C. Várilly: J. Phys. A **21**, L879–L883 (1988)
5. J.C. Várilly, J.M. Gracia-Bondía: Ann. Phys. (NY) **190**, 107–148 (1989)
6. J. Bolte, R. Glaser, S. Keppeler: Ann. Phys. (NY) **293**, 1–14 (2001)
7. J. Liouville: J. Math. Pures Appl. **20**, 137–138 (1855)
8. V.I. Arnold: *Mathematical Methods of Classical Mechanics* (Springer-Verlag, New York, 1978)
9. H. Goldstein: *Classical Mechanics*, 2nd edn. (Addison-Wesley, Reading, Massachusetts, 1980)
10. S. Keppeler: Phys. Rev. Lett. **89**, 210 405 (2002)
11. S. Keppeler: Ann. Phys. (NY) **304**, 40–71 (2003)

5 Torus Quantisation

The old quantum theory was built on the Bohr–Sommerfeld quantisation of classical actions, see [1] and references therein. It was pointed out by Einstein [2] that a necessary condition for the applicability of this method is the integrability of the classical motion and that the quantised objects in phase space are invariant manifolds which are given by tori (cf. Theorem 1).

After the advent of quantum mechanics and, in particular, Schrödinger's wave mechanics, it was shown by Wentzel, Kramers and Brillouin (WKB) [3–5] and also by Jeffreys [6] that the old quantum theory can to a certain extent be derived from the Schrödinger equation in terms of a short wavelength asymptotics. For a WKB-ansatz with a rapidly oscillating phase function it turns out that this phase function has the meaning of a classical action. The quantisation conditions for this action arise when one requires single-valuedness of the wave function. An important modification of the Bohr–Sommerfeld rules (which require the action to be an integer multiple of Planck's constant) comes from classical turning points (or in general from singularities of projections of the underlying Lagrangian manifolds). There the wave function suffers a phase change, which results in a shift of the quantised energies. This shift is nowadays associated with a Maslov index, see [7, 8]. The derivation of the general semiclassical quantisation conditions, based on the geometry of the underlying tori, goes back to Keller [9]. Torus quantisation is therefore often referred to as Einstein–Brillouin–Keller (EBK) quantisation.

Shortly after the development of the relativistic wave equation of the electron [10], Pauli applied the WKB-method to the Dirac equation [11]. He found that the rapidly oscillating phase is given by the action of classical relativistic mechanics and solved the semiclassical equations for a particular example. The role of spin in this treatment was only clarified by Rubinow and Keller [12] who revisited Pauli's ansatz and related the equation which determines the leading order amplitude of the WKB-function to classical spin dynamics given by the equation of Thomas precession [13, 14], see also [15]. However, also in the article [12] no general conditions for torus quantisation are stated, probably due to subtleties that are intrinsic to any semiclassical theory of multi-component wave equations, which in full generality were only to be revealed more than 20 years later.

Influenced by results obtained in [16–18] Littlejohn and Flynn [19, 20] emphasized the importance of geometric phases, see [21, 22], in the semiclassical theory of multicomponent wave equations. They then derived a general torus quantisation scheme for situations in which the (matrix valued) principal symbol of the quantum Hamiltonian has only non-degenerate eigenvalues. This does not cover the case of the Dirac equation for which these eigenvalues (which act as classical Hamiltonians) have multiplicity two. Subsequently Emmrich and Weinstein [23] pointed out that for situations with degenerate eigenvalues integrability of the classical dynamics generated by these eigenvalues is not a sufficient condition to allow for a generalisation of EBK-quantisation.

In Sect. 5.2 we briefly review the WKB-method and explain how it leads to the EBK-quantisation conditions. Then we develop a general torus quantisation scheme for the Pauli and the Dirac equation, see [24, 25], in Sects. 5.3 and 5.5, respectively, under the condition that the underlying classical skew product of translational and spin dynamics is integrable in the sense of Theorem 2. In Sect. 5.6 we show how the newly found quantisation conditions can be applied to the relativistic Kepler problem with spin $\frac{1}{2}$ in order to describe the fine structure of hydrogen. Our results reveal why Sommerfeld [1] was able to derive the correct energy eigenvalues before the development of quantum mechanics and before the discovery of spin: In this particular situation the contribution deriving from the Maslov index and the influence of spin accidentally cancel each other.

Earlier semiclassical treatments of the relativistic hydrogen atom [26–28] where based on a separation of the Dirac equation in spherical coordinates and a subsequent application of the WKB-method to the radial Dirac equation. This approach also reproduces the exact eigenvalues but lacks the clear interpretation of our more general quantisation conditions in which the influence of spin can be identified explicitly. Semiclassical quantisation for subspectra of the Dirac equation based on the complex germ method is discussed in [29, 30]. A semiclassical theory for (matrix valued) Wigner functions of Dirac spinors is described in [31].

5.1 Quantum Mechanical Integrability

Having discussed integrability of classical flows in detail in section 4.4 we can now translate this notion to quantum mechanics. For a scalar system one usually defines quantum mechanical integrability as follows.

Definition 2. *A scalar quantum system with Schrödinger operator $\hat{H}$ is called integrable, if there exists a set of commuting Weyl operators (see Appendix E) $\hat{A}_1 = \hat{H}, \hat{A}_2, \ldots$, i.e. $[\hat{A}_j, \hat{A}_k] = 0\ \forall\ j,k$, whose symbols have an expansion in $\hbar$. In addition their respective principal symbols $A_1, A_2, \ldots$, shall define an integrable classical system in the sense of the requirements listed in Theorem 1.*

It is exactly this class of quantum mechanically integrable systems for which we can find approximate eigenvalues by EBK-quantisation. For the construction of the approximate eigenfunctions one uses most of the properties of classically integrable systems as provided by Theorem 1. Notice that we have not specified the number of commuting observables in Definition 2 since that is already fixed by the requirement that the corresponding classical flow is integrable.

We can now adapt Definition 2 to the case of Pauli operators as follows.

Definition 3. *A Pauli equation is called integrable, if there exists a set of commuting observables $\hat{A}_1 = \hat{H}, \hat{A}_2, \dots$ with Weyl symbols of the form*

$$W[\hat{A}_j](\boldsymbol{p},\boldsymbol{x}) = A_j(\boldsymbol{p},\boldsymbol{x})\,\mathbb{1}_2 + \frac{\hbar}{2}\,\mathrm{d}\pi_s(\boldsymbol{\sigma})\boldsymbol{\mathcal{B}}_j(\boldsymbol{p},\boldsymbol{x}) + \mathcal{O}\left(\hbar^2\right) , \tag{5.1}$$

such that their scalar principal symbols $A_1, A_2, \dots$ define an integrable classical system in the sense of the requirements listed in Theorem 1.

Seemingly, by (5.1) we have imposed additional restrictions on the class of observables $\hat{A}_j$ which we want to consider. However, if we think of observables as functions of position $\boldsymbol{x}$ and momentum $\boldsymbol{p}$ and a spin degree of freedom, then, in a semiclassical context, it is natural to expand this observable for small spin, since spin is proportional to $\hbar$. Thus, our choice of observables (5.1) is rather natural. In addition it is also compatible with our earlier remarks on suitable observables for spinning particles, see Sect. 4.2. Nevertheless, the following treatment can be adapted to the case of arbitrary matrix valued Hamiltonians, but, in order to keep the presentation as simple as possible and physically transparent, we restrict our attention to the situation classified above. We comment on the general case at the end of this chapter in Sect. 5.7.

Now notice that we have not explicitly required condition (4.47) in Definition 3. Therefore, it is not immediately clear that an integrable Pauli equation gives rise to an integrable classical skew product in the semiclassical limit. However, using formula (E.17) for the Weyl symbol of the commutator, we find

$$\begin{aligned} W[[\hat{A}_j,\hat{A}_k]] &= \frac{\hbar}{\mathrm{i}}\{A_j,A_k\} + \frac{\hbar^2}{4}[\mathrm{d}\pi_s(\boldsymbol{\sigma})\boldsymbol{\mathcal{B}}_j, \mathrm{d}\pi_s(\boldsymbol{\sigma})\boldsymbol{\mathcal{B}}_k] \\ &\quad + \frac{\hbar^2}{2\mathrm{i}}\{A_j, \mathrm{d}\pi_s(\boldsymbol{\sigma})\boldsymbol{\mathcal{B}}_k\} + \frac{\hbar^2}{2\mathrm{i}}\{\mathrm{d}\pi_s(\boldsymbol{\sigma})\boldsymbol{\mathcal{B}}_j, A_k\} + \mathcal{O}\left(\hbar^3\right) \\ &= \frac{\hbar}{\mathrm{i}}\{A_j,A_k\} \\ &\quad + \frac{\hbar^2}{2\mathrm{i}}\mathrm{d}\pi_s(\boldsymbol{\sigma})\big(-\boldsymbol{\mathcal{B}}_j \times \boldsymbol{\mathcal{B}}_k + \{A_j,\boldsymbol{\mathcal{B}}_k\} + \{\boldsymbol{\mathcal{B}}_j, A_k\}\big) + \mathcal{O}\left(\hbar^3\right) , \end{aligned} \tag{5.2}$$

i.e. we see that quantum mechanical commutativity does imply property (4.47). Thus, we may use all results of Sect. 4.4.2 when later constructing semiclassical solutions of the Pauli equation.

5.2 EBK-Quantisation

Before we turn our attention to the semiclassical quantisation of integrable systems with spin, let us first summarise some well-known facts about torus or Einstein–Brillouin–Keller (EBK) quantisation for scalar wave equations. A more detailed presentation can, e.g, be found in Keller's original article [9] or in his review article [32].

The goal is to find an approximate solution of the stationary Schrödinger equation

$$\hat{H}\psi(\boldsymbol{x}) = E\psi(\boldsymbol{x}) , \tag{5.3}$$

where $\hat{H}$ is a scalar Weyl operator with principal symbol $H(\boldsymbol{p},\boldsymbol{x})$. In order to simplify the presentation we will assume that in an expansion of the Weyl symbol in powers of $\hbar$ all higher terms vanish. We now make a semiclassical or WKB-ansatz for the wave function of the form

$$\psi(\boldsymbol{x}) = (a_0(\boldsymbol{x}) + \hbar a_1(\boldsymbol{x}) + \ldots)\, \mathrm{e}^{(\mathrm{i}/\hbar)S(\boldsymbol{x})} . \tag{5.4}$$

Upon inserting (5.4) into (5.3) and sorting by powers of $\hbar$ we obtain in leading order the Hamilton–Jacobi equation

$$H(\nabla_{\boldsymbol{x}} S(\boldsymbol{x}), \boldsymbol{x}) = E . \tag{5.5}$$

If $\boldsymbol{X}(t')$ is the spatial part of a solution of Hamilton's equations of motion, the respective momentum part can be written as $\boldsymbol{P}(t') = \nabla_{\boldsymbol{x}} S(\boldsymbol{X}(t'))$. This can be seen by inserting $\boldsymbol{X}(t')$ into (5.5) and differentiating with respect to t'. One finds

$$\nabla_{\boldsymbol{p}} H \frac{\mathrm{d}}{\mathrm{d}t} \nabla_{\boldsymbol{x}} S + \nabla_{\boldsymbol{x}} H \dot{\boldsymbol{X}} = 0 , \tag{5.6}$$

i.e. if $\boldsymbol{X}$ fulfills $\dot{\boldsymbol{X}} = \nabla_{\boldsymbol{p}} H$ then $\nabla_{\boldsymbol{x}} S =: \boldsymbol{P}$ solves $\dot{\boldsymbol{P}} = -\nabla_{\boldsymbol{x}} H$. Thus, for two points $\boldsymbol{y} = \boldsymbol{X}(0)$ and $\boldsymbol{x} = \boldsymbol{X}(t)$ we can write

$$S(\boldsymbol{x}) = S(\boldsymbol{X}(0)) + \int_0^t \left[\frac{\mathrm{d}}{\mathrm{d}t'} S(\boldsymbol{X}(t')) \right] \mathrm{d}t' = S(\boldsymbol{y}) + \int_{\boldsymbol{y}}^{\boldsymbol{x}} \boldsymbol{P}\, \mathrm{d}\boldsymbol{X} , \tag{5.7}$$

where the line integral is along a flow line of ϕ_H^t. In general this rule will not allow us to construct a function $S(\boldsymbol{x})$ which qualifies as the phase of the wave function. For instance, if ϕ_H^t is ergodic on Ω_E, for long times the orbit will return to any small open neighbourhood of $\boldsymbol{y}$ arbitrarily often, but the corresponding values $S(\boldsymbol{x})$ may be completely uncorrelated.

For an integrable system we can proceed as follows. Besides the Hamiltonian $H(\boldsymbol{p},\boldsymbol{x})$ we have $d-1$ further conserved quantities $A_j(\boldsymbol{p},\boldsymbol{x})$, $j = 2,\ldots,d$. Thus, the phase function $S(\boldsymbol{x})$ can be chosen to solve additional Hamilton–Jacobi equations of the form (5.5) with H replaced by A_j, $j = 2,\ldots,d$, and E replaced by other constants. Then by the same construction as above we can obtain $S(\boldsymbol{x})$ by line integration starting from $\boldsymbol{y}$ following the flow lines of the flows ϕ_j^t, $j = 1, 2, \ldots, d$. Since we can reach any point on the joint level set of the observables $H = A_1, A_2, \ldots, A_d$ by some composition of the flows ϕ_j we can define $S(\boldsymbol{x})$ for any desired $\boldsymbol{x}$. Moreover, since $\oint \boldsymbol{P}\mathrm{d}\boldsymbol{X}$ vanishes for every contractible loop, the definition of S does not depend on the order in which we follow the different flow lines. However, recalling that the joint level sets of $A_1, A_2, \ldots, A_d$, are d-tori, there are also non-contractible paths. Thus, by choosing a closed path in (5.7) we find that $S(\boldsymbol{x})$ is in general multi-valued. Any such path is equivalent to a linear combination of paths along the basis cycles C_j. Since the integral

$$\oint_{C_j} \boldsymbol{P}\,\mathrm{d}\boldsymbol{X} = 2\pi I_j \tag{5.8}$$

yields the jth action variable, the values of $S(\boldsymbol{x})$ differ by

$$\Delta S = 2\pi \boldsymbol{k}\boldsymbol{I} \,, \quad \boldsymbol{k} \in \mathbb{Z}^d \,. \tag{5.9}$$

Nevertheless this still allows us to define a single-valued wave function $\psi(\boldsymbol{x})$. For instance, if the amplitude a_0 was single-valued and real, this would demand for the quantisation conditions

$$\exp(2\pi \mathrm{i} I_j/\hbar) = 1 \quad \forall\, j = 1, 2, \ldots, d \quad \Leftrightarrow \quad \boldsymbol{I} = \hbar \boldsymbol{n} \,, \quad \boldsymbol{n} \in \mathbb{Z}^d \,. \tag{5.10}$$

These are the Bohr–Sommerfeld quantisation conditions of the old quantum theory, which – as was pointed out by Einstein [2] – can be applied to any integrable system with no need for separability.

Following Keller [9] we have to modify these conditions due to phase changes occurring in the pre-factor. To this end consider the next-to-leading order equation obtained when inserting ansatz (5.4) in (5.3), the transport equation for $a_0(\boldsymbol{x})$,

$$(\nabla_{\boldsymbol{p}} H)\,(\nabla_{\boldsymbol{x}} a_0) + \frac{1}{2}\,\{\nabla_{\boldsymbol{x}}\,[\nabla_{\boldsymbol{p}} H(\nabla_{\boldsymbol{x}} S(\boldsymbol{x}), \boldsymbol{x})]\}\,a_0 = 0 \,. \tag{5.11}$$

This equation can be solved in much the same way as the respective transport equation which appears in the derivation of the semiclassical time evolution kernel, cf. Appendix B. In general one finds that following a_0 along the paths C_j it suffers a sign change whenever one encounters a caustic. The total sign change along a loop C_j is determined by the Maslov index μ_j of this cycle, see [7, 8]. Thus single-valuedness of ψ now imposes the conditions

$$\begin{aligned} &\frac{2\pi \boldsymbol{I}}{\hbar} - \frac{\pi}{2}\boldsymbol{\mu} = 2\pi \boldsymbol{n} \,, \quad \boldsymbol{n} \in \mathbb{Z}^d \\ \Leftrightarrow \quad &\oint_{C_j} \boldsymbol{P}\,\mathrm{d}\boldsymbol{X} = 2\pi\hbar \left(n_j + \frac{\mu_j}{4} \right) . \end{aligned} \tag{5.12}$$

These are the celebrated EBK quantisation conditions as formulated in this generality by Keller [9].

5.3 Torus Quantisation and Spin Rotation Angles

In this section we give an extension of the EBK quantisation to the case of systems with arbitrary spin. To this end we combine results on the phase space structure of the skew product of translational and spin degrees of freedom (Sect. 4.4.2) with our knowledge about ordinary EBK quantisation (Sect. 5.2).

We are aiming at an approximate solution of the stationary Pauli equation

$$\hat{H}_{\mathrm{P}}\Psi(\boldsymbol{x}) = E\Psi(\boldsymbol{x}) \,, \tag{5.13}$$

where $\Psi \in L^2(\mathbb{R}^d) \otimes \mathbb{C}^{2s+1}$ is a spinor describing a spin-s particle and the Pauli Hamiltonian $\hat{H}_{\mathrm{P}}$ has a matrix valued Weyl symbol,

$$H_{\mathrm{P}}(\boldsymbol{p}, \boldsymbol{x}) = H(\boldsymbol{p}, \boldsymbol{x})\, \mathbb{1}_{2s+1} + \frac{\hbar}{2}\mathrm{d}\pi_s(\boldsymbol{\sigma})\boldsymbol{\mathcal{B}}(\boldsymbol{p}, \boldsymbol{x}) \,. \tag{5.14}$$

Hence, we have a scalar principal symbol and the coupling of spin and translational degrees of freedom is encoded in the sub-principal symbol involving the spin operator

$$\hat{\boldsymbol{s}} = \frac{\hbar}{2}\mathrm{d}\pi_s(\boldsymbol{\sigma}) \tag{5.15}$$

and a field $\boldsymbol{\mathcal{B}}$ which may include terms describing the coupling to an external magnetic field $\boldsymbol{B}(\boldsymbol{x})$ or spin–orbit coupling. The WKB-ansatz of the previous section has to be modified such that in

$$\Psi_{\mathrm{sc}}(\boldsymbol{x}) = (a_0(\boldsymbol{x}) + \hbar a_1(\boldsymbol{x}) + \ldots)\, \mathrm{e}^{(\mathrm{i}/\hbar)S(\boldsymbol{x})} \tag{5.16}$$

the amplitudes $a_k(\boldsymbol{x})$ are now (2s+1)-component spinors; the phase $S(\boldsymbol{x})$ is still scalar. When inserting the ansatz (5.16) into (5.13) the leading order term in $\hbar$ is simply given by the Hamilton–Jacobi equation

$$H(\nabla_{\boldsymbol{x}} S(\boldsymbol{x}), \boldsymbol{x}) = E \,. \tag{5.17}$$

We would have obtained the same equation, had we investigated the stationary Schrödinger equation with Hamiltonian $\hat{H}$ such that its Weyl symbol is $H(\boldsymbol{p}, \boldsymbol{x})$. Thus the whole analysis of the previous section still applies if only the flow ϕ_H^t is integrable in the Liouville–Arnold sense. We find a multivalued phase function $S(\boldsymbol{x})$ by integration along paths on Liouville–Arnold tori. The change in the action $S(\boldsymbol{x})$ after following a basis cycle $\mathcal{C}_j$ is given by the jth action variable,

$$\Delta S = \oint_{\mathcal{C}_j} \boldsymbol{P}\mathrm{d}\boldsymbol{X} = 2\pi I_j \,. \tag{5.18}$$

In the next-to-leading order equation, the transport equation for $a_0(\boldsymbol{x})$, we find an additional term,

$$
\begin{aligned}
(\nabla_{\boldsymbol{p}} H)(\nabla_{\boldsymbol{x}} a_0) + \frac{1}{2}\left\{\nabla_{\boldsymbol{x}}\left[\nabla_{\boldsymbol{p}} H(\nabla_{\boldsymbol{x}} S(\boldsymbol{x}), \boldsymbol{x})\right]\right\} a_0 \\
+ \frac{\mathrm{i}}{2}\, \mathrm{d}\pi_s(\boldsymbol{\sigma}) \boldsymbol{\mathcal{B}}(\nabla_{\boldsymbol{x}} S(\boldsymbol{x}), \boldsymbol{x})\, a_0 = 0\ .
\end{aligned}
\tag{5.19}
$$

However, we may still use the information of the previous section by making the ansatz

$$
a_0(\boldsymbol{x}) = a_0^{\mathrm{S}}(\boldsymbol{x})\, u(\boldsymbol{x})\ , \tag{5.20}
$$

where $a_0^{\mathrm{S}}(\boldsymbol{x})$ is chosen such that it solves the scalar transport equation, i.e. (5.19) with $\boldsymbol{\mathcal{B}} \equiv 0$, and $u(\boldsymbol{x})$ takes values in $\mathbb{C}^{2s+1}$. Hence, $u(\boldsymbol{x})$ has to solve

$$
(\nabla_{\boldsymbol{p}} H)(\nabla_{\boldsymbol{x}} u) + \frac{\mathrm{i}}{2} \mathrm{d}\pi_s(\boldsymbol{\sigma}) \boldsymbol{\mathcal{B}}(\nabla_{\boldsymbol{x}} S(\boldsymbol{x}), \boldsymbol{x})\, u = 0\ . \tag{5.21}
$$

We now have to construct a solution $u(\boldsymbol{x})$ for all points on the respective Liouville–Arnold torus characterised by some set of action variables $\boldsymbol{I}$. Before we tackle this task let us remark that it is a priori not clear whether finding such a solution is possible. In general along a given path on the torus $u(\boldsymbol{x})$ will be multiplied by the representation of some element of SU(2). Choosing two different closed curves we will in general find two different SU(2)-matrices and it is not guaranteed that these commute. If they do not we are prevented from constructing a single-valued WKB-solution for the respective system. This fundamental problem of semiclassics for multi-component wave equations was pointed out in a rather general context by Emmrich and Weinstein [23]. In the present situation we will see that only for systems which are integrable in the sense of Definition 1 we are able to make a suitable choice for $u(\boldsymbol{x})$.

Again we denote by $\boldsymbol{X}(t')$ the spatial part of a solution of Hamilton's equations of motion with $\boldsymbol{X}(0) = \boldsymbol{y}$ and $\boldsymbol{X}(t) = \boldsymbol{x}$ (the corresponding momentum is given by $\boldsymbol{P}(t') = \nabla_{\boldsymbol{x}} S(\boldsymbol{X}(t'))$, cf. Sect. 5.2). Then the first term in (5.21) is a time derivative along the corresponding flow line of ϕ_H^t. Comparing with the spin transport equation (3.43) we conclude that

$$
u(\boldsymbol{x}) = D(\nabla_{\boldsymbol{x}} S(\boldsymbol{y}), \boldsymbol{y}, t)\, u(\boldsymbol{y})\ . \tag{5.22}
$$

If the torus under consideration is ergodic, then for long times a trajectory returns arbitrarily close to $\boldsymbol{y}$ and without further information on the spin dynamics we could not make sure that $u(\boldsymbol{x})$ is single-valued. Even worse than in the similar situation for the action $S(\boldsymbol{x})$, as discussed in Sect. 5.2, we do not only get a multi-valued phase but $u(\boldsymbol{y}) \in \mathbb{C}^{2s+1}$ might have been rotated. Thus, even with a detailed knowledge about this rotation we could in general not compensate it by a suitable quantisation condition, as we were able to make up for the phase changes discussed previously. Hence, integrability of ϕ_H^t is not a sufficient condition to allow for an explicit semiclassical quantisation.

At this point we restrict ourselves to systems which are integrable in the sense of Definition 1, i.e. not only ϕ_H^t but also the skew product Y_{cl}^t is integrable. We have seen that Definition 1 also implies the half classical skew product Y_t, whose spin dynamics still lives on SU(2), to be integrable in the sense that we find d commuting SU(2)-extensions Y_j^t of the flows $\phi_H^t = \phi_1^t, \phi_2^t, \ldots, \phi_d^t$. This allows for a good definition of $u(\boldsymbol{x})$ on the whole torus by

$$u\big((\phi_d^t \circ \cdots \circ \phi_1^t)\boldsymbol{y}\big) = \pi_s[\mathbb{d}(\nabla_{\boldsymbol{x}} S(\boldsymbol{y}), \boldsymbol{y}, \boldsymbol{t})]\, u(\boldsymbol{y}) \,, \tag{5.23}$$

where we have used the notation introduced in (4.61). In this way we can reach any point on the respective torus and the commutativity of the Y_j^t makes it a unique definition up to the contribution of closed loops. These still make $u(\boldsymbol{x})$ multi-valued.

In Sect. 4.4.2 we saw that all SU(2) matrices corresponding to closed loops starting at a given point are confined to a one-parameter subgroup characterised by an axis $\boldsymbol{n}(\boldsymbol{y})$,[1]

$$G_{\boldsymbol{n}(\boldsymbol{y})} = \left\{ g \in \mathrm{SU}(2) \,\middle|\, g = \mathrm{e}^{-(\mathrm{i}/2)\alpha\boldsymbol{\sigma}\boldsymbol{n}(\boldsymbol{y})} \,,\ \alpha \in [0, 4\pi) \right\} . \tag{5.24}$$

We can now choose $u(\boldsymbol{y})$ to be an eigenvector of any $\pi_s(g)$ with $g \in G_{\boldsymbol{n}(\boldsymbol{y})}$. Moving along some path on the torus $u(\boldsymbol{y})$ is multiplied by some $\pi_s(g)$ determined by (5.23). Simultaneously the axis, labeling the local subgroup $G_{\boldsymbol{n}(\boldsymbol{x})}$, is rotated by $\varphi(g) \in \mathrm{SO}(3)$, cf. Sect. 4.4.2. Thus $u(\boldsymbol{x})$ is an eigenvector of any $\pi_s(g)$, $g \in G_{\boldsymbol{n}(\boldsymbol{x})}$ everywhere on the torus. The eigenvalues of $\pi_s(g)$ are given by $\exp(-\mathrm{i}m_s\alpha)$, $m_s = -s, -s+1, \ldots, s$, and to each basis cycle $\mathcal{C}_j$ corresponds a unique rotation angle α_j, cf. Sect. 4.4.2. Hence, the resulting semiclassical solution Ψ_{sc} experiences a total phase change of

$$\frac{2\pi I_j}{\hbar} - \frac{\pi}{2}\mu_j - m_s\alpha_j \tag{5.25}$$

when moving along $\mathcal{C}_j$. The first contribution comes from the multi-valuedness of the action. The second term, the Maslov contribution, derives from the scalar amplitude, and the third term accounts for the spin dynamics. Single-valuedness of the approximate eigenspinor Ψ_{sc} now yields the new quantisation conditions

$$\oint_{\mathcal{C}_j} \boldsymbol{P}\,\mathrm{d}\boldsymbol{X} = 2\pi\hbar \left(n_j + \frac{\mu_j}{4} + m_s\frac{\alpha_j}{2\pi} \right) , \tag{5.26}$$

leading to the semiclassical energy eigenvalues

$$E_{\boldsymbol{n},m_s} = \overline{H}\left(\hbar \left(\boldsymbol{n} + \frac{\boldsymbol{\mu}}{4} + m_s\frac{\boldsymbol{\alpha}}{2\pi} \right) \right) . \tag{5.27}$$

Notice that the scalar case is included in this result with $s = m_s = 0$.

[1] In (4.63) we have indicated a dependence of the axis $\boldsymbol{n}$ on both position and momentum. Since there the phase space points where restricted to d-dimensional tori we may with the same right write $\boldsymbol{n}$ as a function of the coordinates only; locally the corresponding momentum is given by $\nabla_{\boldsymbol{x}} S$.

At this point two remarks concerning the practical implementation of this method are in order. First, applying these quantisation conditions to a particular system, it can be convenient to replace the conserved quantities $A_1, \dots, A_d$ by the action variables $I_1, \dots, I_d$ and complement those to commuting skew products $Y^t_{I_j}$ with fields $\boldsymbol{\mathcal{B}}_{I_j}$. Since the original skew product Y^t_{cl} has to be obtained by a composition of the $Y^t_{I_j}$, where the relative times spent following the different $Y^t_{I_j}$ are determined by the ratios of the frequencies ω_j, i.e.

$$Y^t_{\mathrm{cl}} = Y^{\omega_d t}_{I_d} \circ \dots \circ Y^{\omega_1 t}_{I_1} , \tag{5.28}$$

we have the consistency condition

$$\boldsymbol{\mathcal{B}} = \sum_{j=1}^{d} \omega_j \boldsymbol{\mathcal{B}}_{I_j} . \tag{5.29}$$

The angle α_j is then obtained by investigating

$$\dot{\boldsymbol{s}} = \boldsymbol{\mathcal{B}}_{I_j} \times \boldsymbol{s} , \tag{5.30}$$

where now the dot denotes a time derivative along the flow $\phi^t_{I_j}$, i.e.

$$\dot{\boldsymbol{s}} = \frac{\mathrm{d}\boldsymbol{s}}{\mathrm{d}\vartheta_j} . \tag{5.31}$$

One has to integrate (5.30) from $\vartheta_j = 0$ to 2π and determine the rotation angle α_j. If we are in the favorable situation that $\boldsymbol{\mathcal{B}}_{I_j}$ has a constant direction along this path, we find

$$\alpha_j = \left| \int_0^{2\pi} \boldsymbol{\mathcal{B}}_{I_j} \, \mathrm{d}\vartheta_j \right| . \tag{5.32}$$

The second remark addresses the question of whether the modification of the torus quantisation caused by the spin degrees of freedom can be considered as a small correction to the EBK eigenvalues obtained without spin. Roughly speaking, we ask whether the result can be interpreted as "perturbation theory in $\hbar$". To this end consider the semiclassical eigenvalues (5.27)

$$\begin{aligned} E_{\boldsymbol{n},m_s} &= \overline{H}\Big(\hbar\Big(\boldsymbol{n} + \frac{\boldsymbol{\mu}}{4} + m_s \frac{\boldsymbol{\alpha}}{2\pi}\Big)\Big) \\ &\approx \overline{H}\Big(\hbar\Big(\boldsymbol{n} + \frac{\boldsymbol{\mu}}{4}\Big)\Big) + \boldsymbol{\omega}\Big(\hbar\Big(\boldsymbol{n} + \frac{\boldsymbol{\mu}}{4}\Big)\Big) \, \hbar m_s \frac{\boldsymbol{\alpha}}{2\pi} + \dots , \end{aligned} \tag{5.33}$$

where we have expanded about the result without spin. Due to the trivial appearance of $\hbar$ on the first line one might doubt that this expansion can be interpreted as some kind of semiclassical expansion, although we find an $\hbar$-correction to the EBK energies. However, (5.33) is indeed a semiclassical expansion in the sense of being an expansion for large quantum numbers $\boldsymbol{n}$.

Then classical actions, the argument of the Hamiltonian, are of the order $\mathcal{O}(\hbar|\boldsymbol{n}|) = \mathcal{O}(1)$ and thus large compared to $\hbar$. Therefore, it is legitimate to say that the influence of spin yields an $\hbar$-correction to the semiclassical eigenenergies of the form

$$E_{\text{spin}} = \hbar m_s \frac{\omega\alpha}{2\pi} . \tag{5.34}$$

5.4 Examples

5.4.1 Homogeneous Magnetic Field

As a first example we discuss the situation in which we couple the spin degrees of freedom to a constant magnetic field $\boldsymbol{B}$, i.e. we consider the Pauli Hamiltonian

$$\hat{H}_{\mathrm{P}} = \hat{H} - \frac{e\hbar}{2mc} \mathrm{d}\pi_s(\boldsymbol{\sigma})\boldsymbol{B} , \tag{5.35}$$

where $\hat{H}$ is an arbitrary Schrödinger operator. If E_n is an eigenvalue of $\hat{H}$ then

$$E_n + m_s \frac{e\hbar}{mc}|\boldsymbol{B}| , \quad m_s = -s, \dots, s , \tag{5.36}$$

are eigenvalues of $\hat{H}_{\mathrm{P}}$. Let us see how the semiclassical method finds the respective result.

If we denote by $\overline{H}(\boldsymbol{I})$ the symbol of $\hat{H}$ transformed to action and angle variables, the semiclassical eigenvalues of $\hat{H}$ are given by $\overline{H}(\hbar(\boldsymbol{n} + \boldsymbol{\mu}/4))$, cf. (5.12). Now we can extend the flows generated by the actions $\boldsymbol{I}$ to commuting skew products by setting

$$\boldsymbol{\mathcal{B}}_{I_j} := \frac{e}{mcd\omega_j} \boldsymbol{B} , \tag{5.37}$$

where d is the number of translational degrees of freedom. Clearly, the respective skew products commute, since the $\boldsymbol{\mathcal{B}}_{I_j}$ depend only on the actions $\boldsymbol{I}$ and have a constant direction parallel to $\boldsymbol{B}$. The condition (5.29) is also trivially fulfilled. Now we need to determine the rotation angles for spin transport along a cycle $\mathcal{C}_j$ on a Liouville–Arnold torus. Since the $\boldsymbol{\mathcal{B}}_{I_j}$ are constant on a given torus, the spin rotation angle is given by

$$\alpha_j = 2\pi|\boldsymbol{\mathcal{B}}_{I_j}| = \frac{2\pi}{d\omega_j} \frac{e}{mc}|\boldsymbol{B}| . \tag{5.38}$$

Thus, the equation (5.34) for the energy correction due to spin reads

$$E_{\text{spin}} = \hbar m_s \frac{e|\boldsymbol{B}|}{mc} , \tag{5.39}$$

which is consistent with (5.36).

5.4.2 The σp-Torus

Let us now discuss the $\boldsymbol{\sigma p}$-torus, introduced in Sect. 3.6.3, i.e. a Pauli equation with spin–orbit coupling-term $\kappa(\hbar/2)\mathrm{d}\pi_s(\boldsymbol{\sigma})\hat{\boldsymbol{p}}$ on the rectangle

$$D = \left\{ \boldsymbol{x} \in \mathbb{R}^2 \,|\, 0 \leq x_j \leq a_j \right\} , \tag{5.40}$$

with periodic boundary conditions. For the translational degrees of freedom of the classical system we have already introduced action and angle variables in Sect. 3.6.3 by

$$I_j := \frac{a_j}{2\pi} p_j \quad \text{and} \quad \vartheta_j := \frac{2\pi}{a_j} x_j , \tag{5.41}$$

resulting in the Hamiltonian

$$H(\boldsymbol{I}) = 4\pi^2 \left(\frac{I_1^2}{a_1^2} + \frac{I_2^2}{a_2^2} \right) \quad \text{and frequencies} \quad \omega_j = \frac{8\pi^2 I_j}{a_j^2} . \tag{5.42}$$

One easily verifies that the flows generated by the actions $\boldsymbol{I}$ can be extended to commuting skew products by choosing

$$\boldsymbol{\mathcal{B}}_{I_j} = \kappa \frac{\boldsymbol{p}}{\omega_j d} . \tag{5.43}$$

We then have to solve the following equation of spin precession,

$$\dot{\boldsymbol{s}} = \frac{\kappa}{\omega_j d} \boldsymbol{p} \times \boldsymbol{s} \tag{5.44}$$

on a cycle $\mathcal{C}_j$ along which only ϑ_j changes by 2π. Since both ω_j and $\boldsymbol{p}$ are functions of the actions $\boldsymbol{I}$ only, we immediately find

$$\alpha_j = \kappa \frac{2\pi}{\omega_j d} |\boldsymbol{p}| \tag{5.45}$$

for the rotation angle of a classical spin vector which is transported along $\mathcal{C}_j$. Thus, again using (5.34), we find (as the Maslov indices vanish) the semiclassical eigenvalues

$$E^{\mathrm{sc}}_{\boldsymbol{n},m_s} = 4\pi^2\hbar^2 \left(\frac{n_1^2}{a_1^2} + \frac{n_2^2}{a_2^2} \right) + \kappa m_s \, 2\pi\hbar^2 \sqrt{\frac{n_1^2}{a_1^2} + \frac{n_2^2}{a_2^2}} , \tag{5.46}$$

$m_s = -s, \ldots, s$, which are identical to the exact quantum mechanical ones (3.152).

We remark that the procedure applied here and in the preceeding section is particularly useful in the following class of problems. If the field $\boldsymbol{\mathcal{B}}$, which couples translational and spin degrees of freedom in the Pauli Hamiltonian, expressed as a function of action and angle variables, depends only on the actions $\boldsymbol{I}$ we can always extend the actions to commuting skew products by choosing

$$\boldsymbol{\mathcal{B}}_{I_j}(\boldsymbol{I}) := \frac{\boldsymbol{\mathcal{B}}(\boldsymbol{I})}{\omega_j d} \,. \tag{5.47}$$

The corresponding skew products commute, since $\boldsymbol{\mathcal{B}}_{I_j}(\boldsymbol{I})$ commutes with all $\boldsymbol{I}$ and all fields $\boldsymbol{\mathcal{B}}_{I_j}(\boldsymbol{I})$ are parallel. Since the angles α_j are now trivially found to be $\alpha_j = 2\pi|\boldsymbol{\mathcal{B}}|/(\omega_j d)$, the energy correction (5.34) becomes independent of the frequencies ω_j and we find the semiclassical energies

$$E^{\mathrm{sc}}_{\boldsymbol{n},m_s} = H\left(\hbar\left(\boldsymbol{n} + \boldsymbol{\mu}/4\right)\right) + \hbar m_s \left|\boldsymbol{\mathcal{B}}(\hbar\left(\boldsymbol{n} + \boldsymbol{\mu}/4\right))\right| \,. \tag{5.48}$$

Equations (5.39) and (5.46) are special cases of this result.

5.4.3 Rotationally Invariant Systems

A spin-0 system in 3 dimensions is called rotationally invariant if the Hamiltonian $\hat{H}$ commutes with the operator $\hat{\boldsymbol{L}} := \hat{\boldsymbol{x}} \times \hat{\boldsymbol{p}}$ of angular momentum,

$$[\hat{H}, \hat{\boldsymbol{L}}] = 0 \,. \tag{5.49}$$

We can then choose the eigenfunctions of $\hat{H}$ as simultaneous eigenfunctions of $\hat{H}$, $\hat{\boldsymbol{L}}^2$ and $\hat{L}_z$, since

$$\left[\hat{H}, \hat{\boldsymbol{L}}^2\right] = \left[\hat{H}, \hat{\boldsymbol{L}}_z\right] = \left[\hat{\boldsymbol{L}}^2, \hat{\boldsymbol{L}}_z\right] = 0 \,. \tag{5.50}$$

Similarly, for the classical system whose Hamiltonian is the principal symbol of $\hat{H}$, we also have the commuting observables H, $L := |\boldsymbol{L}|$ and $M := L_z$ ($\boldsymbol{L} := \boldsymbol{x} \times \boldsymbol{p}$),

$$\{H, L\} = \{H, M\} = \{L, M\} = 0. \tag{5.51}$$

Furthermore, the system can be separated in spherical coordinates (r, θ, ϕ), and if we introduce action variables

$$I_r = \oint \frac{\partial S}{\partial r}\,\mathrm{d}r \,, \quad I_\theta = \oint \frac{\partial S}{\partial \theta}\,\mathrm{d}\theta \,, \quad I_\phi = \oint \frac{\partial S}{\partial \phi}\,\mathrm{d}\phi \,, \tag{5.52}$$

where S solves the Hamilton–Jacobi equation we find

$$I_\phi = M \quad \text{and} \quad I_\theta = L - M \,. \tag{5.53}$$

Moreover, in general it can be shown that $\overline{H}(\boldsymbol{I})$ only depends on I_r and the sum $I_\theta + I_\phi$. Therefore, if we choose the set I_r, L and M as our fundamental action variables, the Hamiltonian $\overline{H}$ is independent of M. Correctly determining the Maslov indices one always finds the EBK-quantisation conditions

$$I_r = \hbar\left(n_r + \frac{1}{2}\right) , \quad L = \hbar\left(l + \frac{1}{2}\right) \quad \text{and} \quad M = \hbar m_l \,, \tag{5.54}$$

with integers n_r, l and m_l. From the definitions of L and M as the modulus and z-component of angular momentum, respectively, one always has the constraints

$$L \geq 0 \quad \text{and} \quad |M| \leq L\,. \tag{5.55}$$

This restricts the possible values of the quantum numbers l and m_l to

$$l \in \mathbb{N}_0 \quad \text{and} \quad |m_l| \leq l\,. \tag{5.56}$$

Determining the semiclassical eigenvalues of $\hat{H}$ is thus reduced to finding the Hamiltonian $\overline{H}(I_r, L)$ in action and angle variables.

Let us now see how this scheme translates to the corresponding situation with spin. We call a Pauli operator $\hat{H}_\mathrm{P}$ with spin s rotationally invariant if it commutes with the operator

$$\hat{\boldsymbol{J}} = \hat{\boldsymbol{L}} + \frac{\hbar}{2}\mathrm{d}\pi_s(\boldsymbol{\sigma}) \tag{5.57}$$

of total angular momentum. For the quantum system we can then look for simultaneous eigenfunctions of $\hat{H}_\mathrm{P}$, $\hat{\boldsymbol{J}}^2$ and $\hat{J}_z$ which mutually commute,

$$\left[\hat{H}_\mathrm{P}, \hat{\boldsymbol{J}}^2\right] = \left[\hat{H}_\mathrm{P}, \hat{J}_z\right] = \left[\hat{\boldsymbol{J}}^2, \hat{J}_z\right] = 0\,. \tag{5.58}$$

As previously this has implications for the respective Weyl symbols. We already know the symbols of $\hat{H}_\mathrm{P}$ and $\hat{J}_z$,

$$\begin{aligned} H_\mathrm{P}(\boldsymbol{p}, \boldsymbol{x}) &= H(\boldsymbol{p}, \boldsymbol{x})\,\mathbb{1}_{2s+1} + \frac{\hbar}{2}\mathrm{d}\pi_s(\boldsymbol{\sigma})\boldsymbol{\mathcal{B}}(\boldsymbol{p}, \boldsymbol{x}) \\ J_z &= M\,\mathbb{1}_2 + \frac{\hbar}{2}\mathrm{d}\pi(\sigma_z)\,. \end{aligned} \tag{5.59}$$

For $\boldsymbol{J}^2$ we find

$$\boldsymbol{J}^2 = \boldsymbol{L}^2 + \hbar\,\boldsymbol{L}\,\mathrm{d}\pi_s(\boldsymbol{\sigma}) + \mathcal{O}\left(\hbar^2\right)\,, \tag{5.60}$$

or, since we prefer to work with $|\boldsymbol{J}|$,

$$\begin{aligned} |\boldsymbol{J}| = \sqrt{\boldsymbol{J}^2} &= L + \sqrt{1 + \frac{\hbar \boldsymbol{L}\mathrm{d}\pi_s(\boldsymbol{\sigma})}{L^2} + \mathcal{O}\left(\hbar^2\right)} \\ &= L + \frac{\hbar}{2}\frac{\boldsymbol{L}}{L}\,\mathrm{d}\pi_s(\boldsymbol{\sigma}) + \mathcal{O}\left(\hbar^2\right)\,. \end{aligned} \tag{5.61}$$

Now we can already calculate the modified quantisation conditions for L and M as follows. The z-component M of orbital angular momentum generates a Hamiltonian flow which increases the corresponding angle variable ϑ_M with unit speed. Since the field $\boldsymbol{\mathcal{B}}_M = \boldsymbol{e}_z$, see (5.59), is constant along such a path, the corresponding rotation angle is given by

$$\alpha_M = 2\pi|\boldsymbol{e}_z| = 2\pi\,. \tag{5.62}$$

Thus we find the quantisation condition

$$M = \hbar\,(m_l + m_s) \ , \tag{5.63}$$

$m_l \in \mathbb{Z}$, $m_s = -s, \ldots, s$. Analogously ϕ_L^t increases ϑ_L with unit speed and with $\mathcal{B}_L = \boldsymbol{L}/L$, cf. (5.61), we also find

$$\alpha_L = 2\pi|\mathcal{B}_L| = 2\pi \ . \tag{5.64}$$

Therefore, the modified quantisation condition reads

$$L = \hbar\left(l + \frac{1}{2} + m_s\right) \ , \tag{5.65}$$

$m_l \in \mathbb{Z}$, $m_s = -s, \ldots, s$. The constraints $L \geq 0$ and $|M| \leq L$ now require

$$l + \frac{1}{2} + m_s \geq 0 \quad \text{and} \quad |m_l + m_s| \leq l + \frac{1}{2} + m_s \ , \tag{5.66}$$

for the quantum numbers. Let us rest here for a moment and analyse these conditions, since they look slightly unfamiliar.

As an example consider the case of a spin 1/2. The first condition seems to allow $l = 0$ and $m_s = -1/2$. However, in this situation the second condition reads $|m_l + m_s| \leq 0 \Leftrightarrow m_l = -m_s$ which cannot be fulfilled, since m_l is integer, but m_s is half-integer. Thus, the combination $s = 1/2$, $l = 0$ and $m_s = -1/2$ is forbidden, as expected from exact quantum mechanics. This consideration generalises as follows: We define new semiclassical quantum numbers

$$j := l + m_s \quad \text{and} \quad m_j := m_l + m_s \tag{5.67}$$

for total angular momentum and its z-direction, respectively. If the spin s is integer so is j and if s is half-integer j is also half-integer (since l is always integer). For integer s and j the first condition in (5.66) is equivalent to $j \geq 0$. For half-integer s and j we could also have $j = -1/2$. However, then $L = 0$ and thus the second condition reads $m_l + m_s = 0$ which cannot be fulfilled with integer m_l but half-integer m_s. Thus, we always have the restriction

$$j \geq 0 \ , \tag{5.68}$$

no matter whether $2j$ is even or odd. Furthermore, if s is integer so are m_s and j, and the second condition of (5.66) can be rewritten as $|m_j| \leq j$. On the other hand, if s is half-integer, the left hand side of the second condition is half-integer but the right hand side is integer, i.e. we always have the restriction

$$|m_j| \leq j \ . \tag{5.69}$$

We conclude that we should really interpret j and m_j as defined in (5.67) as quantum numbers characterising the total angular momentum, i.e. the vectorial sum of orbital and spin angular momentum.

What remains to do, in order to determine the semiclassical eigenenergies of a rotationally invariant system with non-zero spin, is to calculate the angle α_r about which a classical spin vector is rotated when transported along a path generated by I_r. The relevant field $\mathcal{B}_r$ can be obtained from condition (5.29) to be

$$\mathcal{B}_r = \frac{\mathcal{B} - \omega_L \mathcal{B}_L - \omega_M \mathcal{B}_M}{\omega_r} = \frac{\mathcal{B} - \omega_L \boldsymbol{L}/L}{\omega_r} \tag{5.70}$$

where we have used that ω_M vanishes and $\mathcal{B}_L$ is defined by (5.61). For a 3-dimensional rotationally invariant system the spin–orbit coupling is proportional to angular momentum and thus both $\mathcal{B}$ and $\mathcal{B}_r$ are parallel to $\boldsymbol{L}$ for all times. Therefore, the equation of spin precession can be integrated as in (5.32) and we find

$$\alpha_r = \left| \int_0^{2\pi} \frac{\mathcal{B} - \omega_L \boldsymbol{L}/L}{\omega_r} \, \mathrm{d}\vartheta_r \right| . \tag{5.71}$$

If we split the integral at the minus sign, the second term is trivial and to solve the first term we can use $\mathrm{d}\vartheta_r/\omega_r = \mathrm{d}\vartheta_r/\dot{\vartheta}_r = \mathrm{d}t$, where t is again the physical time along a trajectory of H. Finally, we obtain

$$\alpha_r = \left| \oint_{\text{radial}} \mathcal{B}(\phi_H^t(\boldsymbol{p}, \boldsymbol{x})) \, \mathrm{d}t - 2\pi \frac{\omega_L}{\omega_r} \frac{\boldsymbol{L}}{L} \right| , \tag{5.72}$$

where the remaining integration is over one cycle of the radial motion, e.g., from perihelion to aphelion and back. We will use this formula when we revisit Sommerfeld's formula for the relativistic hydrogen atom in Sect. 5.6.

5.4.4 Spin–Orbit Coupling in Non-Relativistic Hydrogen

As a first example for a rotationally invariant system we return to the Hamiltonian (3.166) of the non-relativistic Kepler problem with spin–orbit coupling, i.e. the fine structure of hydrogen. It is well-known that the classical Hamiltonian for the Kepler problem expressed in action and angle variables, see e.g [33], reads

$$\overline{H}(\boldsymbol{I}) = -\frac{me^4/2}{(I_r + I_\theta + I_\phi)^2} , \tag{5.73}$$

or, with the definitions of the previous section,

$$\overline{H}(\boldsymbol{I}) = -\frac{me^4/2}{(I_r + L)^2} . \tag{5.74}$$

Here e is the charge of the electron and m its mass. Since I_r and L do not appear independently in (5.74) it is convenient to perform another transformation introducing the action variable

$$I := I_r + L = I_r + I_\theta + I_\phi \,, \tag{5.75}$$

and using the set (I, L, M) to describe the system. Without spin–orbit coupling it is well known that from $I = \hbar n$, $n \in \mathbb{N}$, we obtain the Rydberg formula $E_n = -R/n^2$ with Rydberg energy $R := me^4/(2\hbar^2)$. In order to incorporate spin–orbit coupling we have to determine the rotation angle which corresponds to spin transport along a cycle generated by I. The respective field can be found from (5.29),

$$\boldsymbol{\mathcal{B}}_I = \frac{\boldsymbol{\mathcal{B}}}{\omega} \,, \tag{5.76}$$

with $\boldsymbol{\mathcal{B}} = e^2/(2m^2c^2|\boldsymbol{x}|^3)\boldsymbol{\sigma}\hat{\boldsymbol{L}}$, cf. (3.166), and $\omega = \partial\overline{H}/\partial I = me^4/I^3$. Since $\boldsymbol{\mathcal{B}}_I$ has a constant direction along the relevant cycle (recall that $\boldsymbol{L}$ is a constant of motion) we have

$$\alpha_I = \int_0^{2\pi} \frac{|\boldsymbol{\mathcal{B}}|}{\omega}\, \mathrm{d}\vartheta \,, \tag{5.77}$$

where ϑ denotes the action variable canonically conjugate to I. Now notice that the equation of motion for the Hamiltonian $\overline{H}$ reads

$$\dot{\vartheta} = \frac{\partial\overline{H}}{\partial I} = \omega \,, \tag{5.78}$$

i.e. $\mathrm{d}\vartheta = \omega\, \mathrm{d}t$, where t is the physical time parameterising a trajectory of the original Hamiltonian. Hence we can write

$$\alpha_I = \int_0^T |\boldsymbol{\mathcal{B}}|\, \mathrm{d}t \,, \tag{5.79}$$

where T is the period of the orbit. (Recall that due to the the high degeneracy of the non-relativistic Kepler problem, which is reflected by the fact that $\overline{H}$ depends only on one action variable, all solutions of the equations of motion are periodic.) This integral was already performed in Sect. 3.6.4, cf. (3.170)–(3.174), from which we find

$$\alpha_I = \frac{\pi e^4}{c^2 L^2} \,. \tag{5.80}$$

Therefore we have the modified quantisation condition

$$\begin{aligned} I &= \hbar\left(n + m_s\frac{\alpha_I}{2\pi}\right) = \hbar\left(n + m_s\frac{e^4}{2c^2L^2}\right) \\ &= \hbar\left(n + m_s\frac{e^4}{2c^2\hbar^2(l+\frac{1}{2}+m_s)^2}\right) \end{aligned} \tag{5.81}$$

where we have used (5.65). The quantum numbers take the following values, $n \in \mathbb{N}$, $0 \le l \le n-1$, $m_s = \pm 1/2$ (if $l \ge 1$) or $m_s = 1/2$ (if $l = 0$). Thus we have found semiclassical energies

$$E^{\mathrm{sc}}_{n,m_s} = -\frac{R}{\left(n + \frac{m_s \alpha_{\mathrm{S}}^2}{2(j+\frac{1}{2})^2}\right)^2} , \tag{5.82}$$

where we introduced Sommerfeld's fine structure constant $\alpha_{\mathrm{S}} := e^2/(\hbar c)$ and the quantum number $j = l + m_s$ of total angular momentum.

The result (5.82) is not identical to the energies (3.178) found by resummation of the trace formula in Sect. 3.6.4. However, recall that there the trace formula was not properly derived but, e.g., we used the ad hoc replacement $L \to l + 1/2$. In the light of our discussion in Sect. 5.4.3 we should have rather used $L \to l + 1/2 + m_s$, in which case (3.178) and (5.82) agree.

In many textbooks on quantum mechanics, see e.g. [34, 35], the non-relativistic Kepler problem with spin–orbit coupling is treated in first order perturbation theory in the fine structure constant α_{S}. We can now expand the semiclassical result (5.82) about $\alpha_{\mathrm{S}} = 0$,

$$E^{\mathrm{sc}}_{n,m_s} = -\frac{R}{n^2} + \underbrace{m_s \frac{R\alpha_{\mathrm{S}}^2}{n^3(j+\frac{1}{2})^2}}_{=:\Delta E_{\mathrm{sc}}} + \mathcal{O}\left(\alpha_{\mathrm{S}}^4\right) , \tag{5.83}$$

and compare with the energy correction from perturbation theory which reads

$$\Delta E_{\mathrm{pert}} = \begin{cases} \dfrac{R\alpha_{\mathrm{S}}^2}{2n^3(l+\frac{1}{2})(l+1)} & j = l + \frac{1}{2} \ , \text{ i.e. } m_s = \frac{1}{2} \\ -\dfrac{R\alpha_{\mathrm{S}}^2}{2n^3 l(l+\frac{1}{2})} & j = l - \frac{1}{2} \ , \text{ i.e. } m_s = -\frac{1}{2} \ . \end{cases} \tag{5.84}$$

Both corrections, ΔE_{sc} and ΔE_{pert} agree in leading order for large quantum numbers l (or j, respectively), i.e. in the semiclassical regime where $\hbar$ is small compared to the action variable L.

A third non-relativistic expression for the energy levels of a hydrogen atom with spin–orbit coupling can be obtained by solving the exact relativistic problem, i.e. the Dirac equation with Coulomb potential, and expanding the result – the so-called Sommerfeld formula, see also Sect. 5.6 – in powers of $1/c$ or, equivalently, α_{S}. However, one neither finds agreement with ΔE_{sc} nor with ΔE_{pert}, as in the same order for $c \to \infty$ further corrections contribute, e.g., a term proportional to $\hat{\boldsymbol{p}}^4$ which derives from the relativistic mass. Nevertheless, we remark that ΔE_{sc} correctly reflects the symmetry of the problem, in the sense that the quantum number l of orbital angular momentum appears only in combination with m_s such that the result can be nicely written in terms of the quantum number j of total angular momentum. The latter enters in the same way as it does in the Sommerfeld formula, via terms which are functions of $(j + 1/2)$.

5.5 Spin Rotation Angles in the Dirac Case

We briefly show how the reasoning of the preceeding sections, i.e. the construction of approximate semiclassical solutions for the Pauli equation, can be extended to the case of the Dirac equation. This will give us the opportunity to revisit Sommerfeld's quantisation of the relativistic Kepler problem, i.e. the fine structure of the hydrogen atom, in the following section. In order to keep the presentation short we heavily rely on the analogy to the derivation in Sect. 5.3. More details on the relativistic case an be be found in [24, 25].

We intend to find semiclassical solutions of the Dirac equation,

$$\hat{H}_{\mathrm{D}} \Psi(\boldsymbol{x}) = E \Psi(\boldsymbol{x}) , \tag{5.85}$$

with Dirac Hamiltonian (3.183). The semiclassical ansatz reads

$$\Psi(\boldsymbol{x}) = \left(a_0(\boldsymbol{x}) + \hbar a_1(\boldsymbol{x}) + \dots\right) \mathrm{e}^{(\mathrm{i}/\hbar) S(\boldsymbol{x})} , \tag{5.86}$$

where the amplitudes $a_k(\boldsymbol{x})$ are four-spinors, i.e. they take values in $\mathbb{C}^4$, and the phase $S(\boldsymbol{x})$ is scalar. Inserting this ansatz and sorting by like orders in $\hbar$, in leading order we find

$$\left[H_{\mathrm{D}}(\nabla S(\boldsymbol{x}), \boldsymbol{x}) - E\right] a_0(\boldsymbol{x}) = 0 , \tag{5.87}$$

where H_{D} denotes the Weyl symbol (3.185) of the Dirac Hamiltonian $\hat{H}_{\mathrm{D}}$. Recalling the discussion following equation (3.192) we know that equation (5.87) can only be solved if $S(\boldsymbol{x})$ fulfills either of the two Hamilton–Jacobi equations

$$H^{\pm}(\nabla S^{\pm}(\boldsymbol{x}), \boldsymbol{x}) = E \tag{5.88}$$

with Hamiltonians

$$H^{\pm}(\boldsymbol{p}, \boldsymbol{x}) = e\phi \pm \sqrt{c^2 \left(\boldsymbol{p} - \frac{e}{c}\boldsymbol{A}\right)^2 + m^2 c^4} . \tag{5.89}$$

Simultaneously, the leading term of the amplitude must be of the form

$$a_0^{\pm}(\boldsymbol{x}) = V_{\pm}(\nabla S^{\pm}(\boldsymbol{x}), \boldsymbol{x})\, b_{\pm}(\boldsymbol{x}) , \tag{5.90}$$

where $b_{\pm}(\boldsymbol{x})$ is a two-spinor, i.e. $b_{\pm} \in \mathbb{C}^2$, and the 4×2 matrices $V_{\pm}(\boldsymbol{p}, \boldsymbol{x})$ were defined in (3.187). An equation for $b_{\pm}$ is obtained by multiplying the next-to-leading order equation,

$$\left[H_{\mathrm{D}}(\nabla S^{\pm}, \boldsymbol{x}) - E\right] a_1^{\pm} + c\boldsymbol{\alpha}\nabla\, a_0^{\pm} = 0 , \tag{5.91}$$

from the left with $V_{\pm}^{\dagger}(\nabla S^{\pm}(\boldsymbol{x}), \boldsymbol{x})$. Then the same calculation which led to (3.198) yields,

$$\begin{aligned} &\nabla_{\boldsymbol{p}} H^{\pm}(\nabla S^{\pm}, \boldsymbol{x})\, (\nabla b_{\pm}) \\ &+ \left[\frac{1}{2}\nabla_{\boldsymbol{x}} \left(\nabla_{\boldsymbol{p}} H^{\pm}(\nabla_{\boldsymbol{x}} S^{\pm}, \boldsymbol{x})\right) + \frac{\mathrm{i}}{2}\boldsymbol{\mathcal{B}}^{\pm}(\nabla_{\boldsymbol{x}} S^{\pm}, \boldsymbol{x})\, \boldsymbol{\sigma}\right] b_{\pm} = 0 . \end{aligned} \tag{5.92}$$

The fields $\mathcal{B}^{\pm}$ are given by

$$\mathcal{B}^{\pm}(\boldsymbol{p},\boldsymbol{x}) = \frac{ec^2}{\epsilon(\epsilon+mc^2)}\left[\left(\boldsymbol{p}-\frac{e}{c}\boldsymbol{A}\right)\times\boldsymbol{E}\right] \mp \frac{ec}{\epsilon}\boldsymbol{B}\,. \tag{5.93}$$

see (3.199). They account for spin–orbit coupling and coupling of spin to the external magnetic field $\boldsymbol{B}$. Notice that (5.92) is of the same form as (5.19). Therefore, using the construction of Sect. 5.3 we can immediately state the result. If the skew product flows $Y_{\pm}^t$ with base flows $\phi_{\pm}^t$ generated by the Hamiltonians (5.89) and spin precession

$$\dot{s} = \mathcal{B}^{\pm}\times s \tag{5.94}$$

are integrable in the sense of Definition 1, we can find semiclassical solutions of the Dirac equation with approximate energy eigenvalues

$$E^{\pm}_{\boldsymbol{n},m_s} = \overline{H}^{\pm}\left(\hbar\left(\boldsymbol{n}+\frac{\boldsymbol{\mu}}{4}+m_s\frac{\boldsymbol{\alpha}}{2\pi}\right)\right)\,. \tag{5.95}$$

Here $\overline{H}^{\pm}(\boldsymbol{I})$ are the Hamiltonians $H^{\pm}(\boldsymbol{p},\boldsymbol{x})$ expressed in action and angle variables $(\boldsymbol{I},\boldsymbol{\vartheta})$, n_j are integers, μ_j is the Maslov index corresponding to the jth basis cycle $\mathcal{C}_j$ on the respective Liouville–Arnold torus of $\phi_{\pm}^t$, and α_j is the respective rotation angle for a classical spinor transported along $\mathcal{C}_j$ as explained in Sects. 4.4.2 and 5.3. The example which we will investigate in the following section will nicely illustrate the procedure and the different quantities.

5.6 The Sommerfeld Formula

In 1916 Arnold Sommerfeld calculated the fine structure of the hydrogen atom [1]. He used a relativistic treatment of the Kepler problem which he quantised with the conditions (5.10) of the "old" quantum theory, nowadays referred to as the Bohr–Sommerfeld quantisation conditions. At that time quantum mechanics was not yet developed, neither the matrix mechanics of Born, Heisenberg and Jordan nor Schrödinger's wave mechanics – not to mention relativistic wave equations. He also could have no knowledge of the spin of the electron, but yet he obtained the correct energy levels of the hydrogen atom. More than 10 years later, after Dirac had developed the quantum mechanics of relativistic particles with spin 1/2 [10], it turned out [36, 37] that the Dirac theory yields exactly the same eigenvalues as Sommerfeld's treatment of the problem. In the textbook "Atombau und Spektrallinien" [38] Sommerfeld later writes that it is still instructive to do the calculation with the old method first, since the results are the same and because the new method "is not only less intuitive but also much more complicated":

> *"Wir behandeln das relativistische Keplerproblem zunächst elementar, gehen aber alsbald (...) zur allgemeinen 'Hamilton-Methode' über, die der Natur unseres Problems merkwürdig konform ist. (...)*
> *Die im vorstehenden berechneten Energie-Niveaus und die daraus abzuleitenden Liniengebilde behalten auch wellenmechanisch exakte Gültigkeit. Der Weg, den die Wellenmechanik zu ihrer Ableitung einschlagen muß, ist nicht nur viel weniger anschaulich sondern auch viel umständlicher als der hier begangene Weg. Deshalb läßt es sich rechtfertigen, die Energieberechnung zunächst nach der Methode der alten Quantentheorie durchzuführen; die daraus im folgenden zu ziehenden Konsequenzen können später direkt in die Wellenmechanik übernommen werden."* [38, pp. 273 & 279]

By treating the problem with the semiclassical method developed in the preceeding sections we shed some light on the amazing success of Sommerfeld's result.

We start by listing some properties of the relativistic Kepler problem. Details can, e.g., be found in Sommerfeld's book [38]. The Hamiltonian of the relativistic Kepler problem is given by

$$H(\boldsymbol{p}, \boldsymbol{x}) = -\frac{e^2}{|\boldsymbol{x}|} + \sqrt{c^2\boldsymbol{p}^2 + m^2c^4} \,. \tag{5.96}$$

Transformed to action and angle variables it reads

$$\overline{H}(I_r, L) = mc^2 \left[1 + \frac{e^4/c^2}{\left(I_r + \sqrt{L^2 - e^4/c^2}\right)^2} \right]^{-1/2} , \tag{5.97}$$

where we used the notation of Sect. 5.4.3, since, obviously, the problem is rotationally invariant. The solutions of the equations of motion are "Rosettenbahnen", ellipses with moving perihelia. Expressed in the xy-plane (i.e. $\theta = \frac{\pi}{2}$) the orbits are given by

$$\frac{1}{r(\phi)} = \underbrace{\frac{e^2E}{c^2L^2 - e^4}}_{:=C} + \underbrace{\frac{\sqrt{c^2L^2E^2 + (c^2L^2 - e^4)m^2c^4}}{c^2L^2 - e^4}}_{:=A} \cos\left(\underbrace{\frac{\sqrt{c^2L^2 - e^4}}{cL}}_{:=\gamma} \phi \right), \tag{5.98}$$

where we have chosen the orientation of the orbit such that the perihelion is at $\phi = 0$. For later reference we introduced the abbreviations C, A and γ.

Sommerfeld quantised the system by choosing [1]

$$I_r = \hbar n_r \quad \text{and} \quad L = \hbar l \,, \tag{5.99}$$

with integers $n_r \in \mathbb{N}_0$ and $l \in \mathbb{N}$. Comparing with the exact eigenvalues of the Dirac equation, see e.g. [34, 35, 39], one finds that $\overline{H}(\hbar n_r, \hbar l)$ yields the correct energy levels. With today's knowledge of semiclassical quantisation, even neglecting spin, we would say that we should include the Maslov contribution which changes the quantisation conditions to

$$I_r = \hbar \left(n_r + \frac{1}{2} \right) \quad \text{and} \quad L = \hbar \left(l + \frac{1}{2} \right) . \tag{5.100}$$

Then the semiclassical energies no longer coincide with those obtained from the Dirac equation. Instead they are identical to the eigenenergies found using the Klein–Gordon equation, the relativistic wave equation for particles with spin 0. This makes sense, since by applying the EBK-quantisation scheme to the Klein–Gordon equation for a "spinless hydrogen" we arrive at the same semiclassical quantisation.

Now we go one step further and also include the spin contribution as derived in the preceeding sections. To this end we have to "quantise" the skew product Y^t_{cl} which consists of the translational dynamics with Hamiltonian (5.96) and the spin precession

$$\dot{\boldsymbol{s}} = \boldsymbol{\mathcal{B}} \times \boldsymbol{s} . \tag{5.101}$$

The effective field $\boldsymbol{\mathcal{B}}$ is obtained from $\boldsymbol{\mathcal{B}}^+$ of equation (3.199) by inserting $\boldsymbol{E} = -\nabla(-e^2/r) = -e^2 \boldsymbol{x}/r^3$, i.e.

$$\boldsymbol{\mathcal{B}} = \frac{e^2 c^2}{\epsilon(\epsilon + mc^2)} \frac{1}{r^3} \boldsymbol{L} . \tag{5.102}$$

Recall that ϵ is the kinetic energy, $\epsilon = \sqrt{c^2 \boldsymbol{p}^2 + m^2 c^4}$, defined to include the rest energy. Without loss of generality we may consider the equation of spin precession for an orbit in the xy-plane, in which case

$$\boldsymbol{L} = \frac{\epsilon}{c^2} r^2 \dot{\phi} \, \boldsymbol{e}_z , \tag{5.103}$$

i.e.

$$\boldsymbol{\mathcal{B}} = \frac{e^2}{\epsilon + mc^2} \frac{1}{r} \dot{\phi} \, \boldsymbol{e}_z . \tag{5.104}$$

Moreover, we can use (5.96) and $H(\boldsymbol{p}, \boldsymbol{x}) = E$ to write

$$\boldsymbol{\mathcal{B}} = \frac{e^2}{(E + mc^2) r + e^2} \dot{\phi} \, \boldsymbol{e}_z . \tag{5.105}$$

Now we have assembled all the ingredients needed in order to perform the programme of Sect. 5.3. Since the problem is rotationally invariant we can rely on the results of Sect. 5.4.3. Thus, we immediately obtain the quantisation condition for the angular momentum L,

$$L = \hbar\left(l + \frac{1}{2} + m_s\right), \quad l \in \mathbb{N}_0\,, \; m_s = \pm\frac{1}{2}\,, \tag{5.106}$$

see equation (5.65). Recalling that the combination $l = 0$, $m_s = -1/2$ is forbidden, see the discussion below (5.55), we see that (5.106) yields the same values $L/\hbar \in \mathbb{N}$ as Sommerfeld's condition (5.99). In order to also quantise the radial action I_r we need to calculate the angle α_r by which a spin vector is rotated when transported along a cycle generated by I_r. The general formula for rotationally invariant systems was derived in Sect. 5.4.3 and reads, see (5.72),

$$\alpha_r = \left| \oint_{\text{radial}} \boldsymbol{\mathcal{B}}\,\mathrm{d}t - 2\pi\,\frac{\omega_L}{\omega_r}\,\frac{\boldsymbol{L}}{L} \right| . \tag{5.107}$$

With (5.97) we find

$$\frac{\omega_L}{\omega_r} = \frac{\partial \overline{H}/\partial L}{\partial \overline{H}/\partial I_r} = \frac{\partial}{\partial L}\sqrt{L^2 + \frac{e^4}{c^2}} = \frac{cL}{\sqrt{c^2L^2 + e^4}} = \frac{1}{\gamma}\,, \tag{5.108}$$

and with (5.105) we have to calculate the integral

$$\boldsymbol{\mathcal{I}} := \oint_{\text{radial}} \frac{e^2}{(E + mc^2)\,r(\phi) + e^2}\,\boldsymbol{e}_z\,\mathrm{d}\phi \tag{5.109}$$

where we have to insert $r(\phi)$ from (5.98) and the ϕ-integration goes over one cycle of the radial motion, e.g., from perihelion to perihelion., i.e. $0 \le \phi \le 2\pi/\gamma$. Changing variables to $\eta := \gamma\phi$ the integral reads

$$\begin{aligned}
\boldsymbol{\mathcal{I}} &= \frac{\boldsymbol{e}_z}{\gamma}\int_0^{2\pi} \frac{e^2}{\frac{E+mc^2}{C+A\cos(\eta)} + e^2}\,\mathrm{d}\eta \\
&= \frac{\boldsymbol{e}_z}{\gamma}\int_0^{2\pi} \frac{e^2C + e^2A\cos(\eta)}{E + mc^2 + e^2C + e^2A\cos(\eta)}\,\mathrm{d}\eta \\
&= \frac{\boldsymbol{e}_z}{\gamma}\int_0^{2\pi} \left[1 - \frac{E + mc^2}{E + mc^2 + e^2C + e^2A\cos(\eta)}\right]\mathrm{d}\eta \\
&= \frac{2\pi\boldsymbol{e}_z}{\gamma} - \frac{\boldsymbol{e}_z}{\gamma}\underbrace{\int_0^{2\pi} \frac{E + mc^2}{E + mc^2 + e^2C + e^2A\cos(\eta)}\,\mathrm{d}\eta}_{2\pi\gamma}\,,
\end{aligned} \tag{5.110}$$

where the last integral can be calculated with standard methods. Inserting (5.108) and (5.110) into (5.107), and recalling that we have chosen $\boldsymbol{L} \| \boldsymbol{e}_z$, we find

$$\alpha_r = \left| \frac{2\pi}{\gamma}\boldsymbol{e}_z + 2\pi\boldsymbol{e}_z - \frac{2\pi}{\gamma}\frac{\boldsymbol{L}}{L} \right| = 2\pi\,. \tag{5.111}$$

Therefore the correct semiclassical quantisation condition for the radial action reads

$$I_r = \hbar\left(n_r + \frac{1}{2} + m_s\right) \tag{5.112}$$

with n_r integer and $m_s = \pm\frac{1}{2}$. Thus, the radial action is also quantised by integer multiples of $\hbar$ and with $I_r \geq 0$ (a condition which comes from classical mechanics) we find that $I_r/\hbar \in \mathbb{N}_0$ again takes the same values as in Sommerfeld's condition (5.99). We conclude that the semiclassical energies

$$E_{n_r,l,m_s} = mc^2\left[1 + \frac{\alpha_S^2}{\left(n_r + \frac{1}{2} + m_s + \sqrt{\left(l + \frac{1}{2} + m_s\right)^2 - \alpha_S^2}\right)^2}\right]^{-1/2}, \tag{5.113}$$

where $\alpha_S = e^2/\hbar c$ is Sommerfeld's fine structure constant, are identical to those obtained by Sommerfeld, because, roughly speaking, the Maslov contribution and the correction which comes from the spin of the electron cancel each other. Funnily enough, a similar cancellation also occurred in our introductory example, the Dirac oscillator (see Sect. 2.1). Certainly, Sommerfeld could not obtain the correct multiplicities for the eigenvalues, since his quantum numbers are combinations of the 'real' quantum numbers. This problem is rectified by the present treatment, where the multiplicities are identical to those obtained from the full Dirac theory [36, 37].

We remark that the result can be put in a more familiar form by introducing the quantum number $j := l + m_s$ of total angular momentum as in (5.67) and the so-called principal quantum number $n \in \mathbb{N}$ associated with the action variable $I = I_r + L$ by $I := \hbar n$, i.e.

$$n = n_r + l + 1 + 2m_s = n_r + j + 1 + m_s\ . \tag{5.114}$$

Then the energy eigenvalues (5.113) read

$$E_{n,j} = mc^2\left[1 + \frac{\alpha_S^2}{\left(n - \left(j + \frac{1}{2}\right) + \sqrt{\left(j + \frac{1}{2}\right)^2 - \alpha_S^2}\right)^2}\right]^{-1/2}, \tag{5.115}$$

see e.g. [39].

5.7 Excursion: Remarks on the General Case

After discussing semiclassical quantisation for the Pauli and the Dirac equation in detail, let us finally make some remarks on the general case of arbitrary matrix valued Hamiltonians $\hat{H}$ acting on $L^2(\mathbb{R}^d) \otimes \mathbb{C}^n$.

In general the matrix valued (principal) Weyl symbol $H(\boldsymbol{p},\boldsymbol{x})$ of $\hat{H}$ will have different eigenvalues $H_1(\boldsymbol{p},\boldsymbol{x}),\dots,H_m(\boldsymbol{p},\boldsymbol{x})$, $m \le n$. The multiplicity of each eigenvalue shall be given by $g_j \in \mathbb{N}$, $j=1,\dots,m$ with $\sum_j g_j = n$. We will not allow the multiplicities to change when varying $(\boldsymbol{p},\boldsymbol{x})$, i.e. we exclude the problem of mode conversion, see [40], for a discussion. Say we want to find semiclassical energy eigenvalues associated with the classical dynamics generated by $H_j(\boldsymbol{p},\boldsymbol{x})$. Then we have to project the transport equation for the leading term of the amplitude of a semiclassical eigenvector onto the subspace of $\mathbb{C}^n$ corresponding to $H_j(\boldsymbol{p},\boldsymbol{x})$ in exactly the same way as we did for the Dirac equation with classical Hamiltonian $H^{\pm}(\boldsymbol{p},\boldsymbol{x})$. The classical flow generated by $H_j(\boldsymbol{p},\boldsymbol{x})$ together with the projected transport equation gives rise to a skew product Y^t, which in general is a $\mathrm{U}(g_j)$-extension of $\phi^t_{H_j}$. We shall call Y^t integrable if

1. $\phi^t_{H_j}$ is integrable in the sense of Liouville–Arnold, see Theorem 1 and if
2. we can define $\mathrm{U}(g_j)$-extensions of the flows $\phi^t_2,\dots,\phi^t_d$ which all commute with each other and with Y^t, cf. [25, Proposition 5].

Provided that Y^t is integrable consider a Liouville–Arnold torus of $\mathbb{T}^d$ of $\phi^t_{H_j}$. To each basis cycle $\mathcal{C}_1,\dots,\mathcal{C}_d$ starting at a given point on $\mathbb{T}^d$ we can associate a $d_l \in \mathrm{U}(g_j)$, $l=1,\dots,d$, by the same procedure as in Sect. 5.3, where instead of $\mathrm{U}(g_j)$ we had to consider $\mathrm{SU}(2)$. $\mathrm{U}(g_j)$-elements d'_l associated with a cycle starting at a different point on $\mathbb{T}^d$ are obtained by conjugation with a group element $h \in \mathrm{U}(g_j)$ independent of l, i.e. $d'_l = h d_l h^{-1}$. Moreover, commutativity of the $\mathrm{U}(g_j)$-extensions of $\phi^t_{H_j},\phi^t_2,\dots,\phi^t_d$ guarantees that all d_l, $l=1,\dots,d$, commute. Thus, we can choose a basis in $\mathbb{C}^{g_j}$ such that all d_l are diagonalised simultaneously with eigenvalues

$$\mathrm{e}^{\mathrm{i}\varphi_{l,\lambda}}\,, \tag{5.116}$$

where the index $l=1,\dots,d$ labels the corresponding cycle $\mathcal{C}_l$ and $\lambda = 1,\dots,g_j$ labels the direction in $\mathbb{C}^{g_j}$. The phases $\varphi_{l,\lambda}$ play the same role as the terms $m_s\alpha$ in the case where instead of $\mathrm{U}(g_j)$ we had a $(2s+1)$-dimensional representation of $\mathrm{SU}(2)$. Thus, the semiclassical quantisation conditions now read

$$\oint_{\mathcal{C}_l} \boldsymbol{p}\,\mathrm{d}\boldsymbol{x} = 2\pi\hbar\left(n_l + \frac{\mu_l}{4} + \frac{\varphi_{l,\lambda}}{2\pi}\right)\,, \tag{5.117}$$

where n_l is an integer and μ_l denotes the Maslov index of the cycle $\mathcal{C}_l$. The index $\lambda = 1,\dots,g_j$ plays the role of the spin quantum number $m_s=-s,\dots,s$ in the previous cases.

References

1. A. Sommerfeld: Ann. Phys. (Leipzig) **51**, 1–94, 125–167 (1916)
2. A. Einstein: Verh. Dtsch. Phys. Ges. **19**, 82–92 (1917)

3. G. Wentzel: Z. Physik **38**, 519–529 (1926)
4. H.A. Kramers: Z. Physik **39**, 828–840 (1926)
5. L. Brillouin: Compt. Rend. **183**, 24–26 (1926)
6. H. Jeffreys: Proc. London Math. Soc. **23**, 428–436 (1925)
7. V.P. Maslov: *Théorie des Perturbations et Méthodes Asymptotiques* (Dunod, Paris, 1972)
8. V.P. Maslov, M.V. Fedoriuk: *Semi-Classical Approximation in Quantum Mechanics* (D. Reidel, Dodrecht, 1981)
9. J.B. Keller: Ann. Phys. (NY) **4**, 180–185 (1958)
10. P.A.M. Dirac: Proc. R. Soc. London Ser. A **117**, 610–624 (1928)
11. W. Pauli: Helv. Phys. Acta **5**, 179–199 (1932)
12. S.I. Rubinow, J.B. Keller: Phys. Rev. **131**, 2789–2796 (1963)
13. L.H. Thomas: Nature **117**, 514 (1926)
14. L.H. Thomas: Philos. Mag. **3**, 1–22 (1927)
15. V. Bargman, L. Michel, V.L. Telegdi: Phys. Rev. Lett. **2**, 435–436 (1959)
16. K. Yabana, H. Horiuchi: Prog. Theor. Phys. **77**, 517–547 (1987)
17. H. Kuratsuji, S. Iida: Prog. Theor. Phys. **74**, 439–445 (1985)
18. H. Kuratsuji, S. Iida: Phys. Rev. D **37**, 441–447 (1988)
19. R.G. Littlejohn, W.G. Flynn: Phys. Rev. Lett. **66**, 2839–2842 (1991)
20. R.G. Littlejohn, W.G. Flynn: Phys. Rev. A **44**, 5239–5256 (1991)
21. M.V. Berry: Proc. R. Soc. London Ser. A **392**, 45–57 (1984)
22. A. Shapere, F. Wilczek: *Geometric Phases in Physics* (World Scientific Publishing, Singapore, 1989)
23. C. Emmrich, A. Weinstein: Commun. Math. Phys. **176**, 701–711 (1996)
24. S. Keppeler: Phys. Rev. Lett. **89**, 210 405 (2002)
25. S. Keppeler: Ann. Phys. (NY) **304**, 40–71 (2003)
26. R.H. Good: Phys. Rev. **90**, 131–137 (1953)
27. M. Rosen, D.R. Yennie: J. Math. Phys. **5**, 1505–1515 (1964)
28. P. Lu: Phys. Rev. A **1**, 1283–1285 (1970)
29. V.G. Bagrov, V.V. Belov, A.Yu. Trivonov, A.A. Yevseyevich: J. Phys. A **27**, 1021–1043 (1994)
30. V.G. Bagrov, V.V. Belov, A.Yu. Trivonov, A.A. Yevseyevich: J. Phys. A **27**, 5273–5306 (1994)
31. H. Spohn: Ann. Phys. (NY) **282**, 420–431 (2000)
32. J.B. Keller: SIAM Rev. **27**, 485–504 (1985)
33. H. Goldstein: *Classical Mechanics*, 2nd edn. (Addison-Wesley, Reading, Massachusetts, 1980)
34. A. Messiah: *Quantum Mechanics. Vol. II*, Translated from the French by J. Potter (North-Holland Publishing Co., Amsterdam, 1962)
35. P. Strange: *Relativistic Quantum Mechanics with Applications in Condensed Matter and Atomic Physics* (Cambridge University Press, Cambridge, 1998)
36. W. Gordon: Z. Phys. **48**, 11–14 (1928)
37. C.G. Darwin: Proc. R. Soc. London Ser. A **118**, 654–680 (1928)
38. A. Sommerfeld: *Atombau und Spektrallinien*, Vol. II, 4th edn. (Friedr. Viehweg & Sohn GmbH, Braunschweig, 1967)
39. J.D. Bjorken, S.D. Drell: *Relativistic Quantum Mechanics* (McGraw-Hill, New York, St. Louis, San Francisco, 1964)
40. W.G. Flynn, R.G. Littlejohn: Ann. Physics (NY) **234**, 334–403 (1994)

6 Classical Sum Rules

In this chapter we discuss classical sum rules which relate certain periodic orbit sums to ergodic averages. These sum rules can then be used in order to obtain semiclassical estimates for the quantum mechanical spectral two-point form factor.

The physical picture behind a classical sum rule is the following. Considering an ergodic flow we may replace phase space averages by time averages along typical trajectories. On the other hand, a typical trajectory will locally be close to some periodic trajectory (since in typical systems these form a dense subset). Thus, we might expect that the average along the typical trajectory can be replaced by an average along many periodic orbits. However, not all these periodic orbits will contribute with the same weight. The time a typical orbits stays in the vicinity of a periodic orbit depends on its stability and thus the contributions to the periodic orbit sum are, amongst other things, weighted by the stability. This is reminiscent of the structure of semiclassical trace formulae. As a matter of fact one finds that the amplitudes in classical sum rules are the absolute value squared of the amplitudes in semiclassical trace formulae. Thus, it is immediately clear why these sum rules are important tools for a semiclassical theory of spectral two-point correlations.

The concept of classical sum rules was developed in a seminal article by Hannay and Ozorio de Almeida [1], for which reason these relations are also known under the notion of Hannay-Ozorio de Almeida sum rules. Equivalent results in the case of hyperbolic dynamics are known in the mathematical literature in the context thermodynamic formalism, see [2] and references therein. Classical sum rules are also intimately related to so-called classical trace formulae, see [3, 4]

We informally elaborate on the basic idea behind classical sum rules and how to derive them in Sect. 6.1. A brief account of some mathematical subtleties involved is given in Sect. 6.2. In Sect. 6.3 we review the derivation of Hannay-Ozorio de Almeida sum rules for hyperbolic and integrable systems. After discussing suitable classical time evolution operators for the skew product dynamics of translational and spin degrees of freedom and the spaces they should naturally be defined on in Sect. 6.4, we turn to the derivation of classical sum rules for systems with spin in Sect. 6.5, where we distinguish three situations: Hyperbolic translational dynamics with an ergodic skew product, integrable skew products (in the sense of Theorem 2), and a partially integrable case with integrable translational dynamics but a non-integrable skew product.

6.1 Basic Idea

In order to illustrate the basic idea behind classical sum rules consider a flow ϕ^t on a manifold $\mathcal{M}$ preserving some normalised measure μ. On $L^2(\mathcal{M})$ we define a time evolution operator U^t by

$$U^t f(z) = f(\phi^t(z)) , \quad \forall f \in L^2(\mathcal{M}) , \tag{6.1}$$

which we can be represented in terms of an integral kernel $K_{\text{cl}}(z, z', t)$, i.e.

$$U^t f(z) = \int_{\mathcal{M}} K_{\text{cl}}(z, z', t) f(z') \, \mathrm{d}\mu(z') , \tag{6.2}$$

with the choice

$$K_{\text{cl}}(z, z', t) = \delta_{\mathcal{M}}(\phi^t(z) - z') . \tag{6.3}$$

Here the δ-function is defined such that

$$\int_{\mathcal{M}} f(z') \, \delta_{\mathcal{M}}(z - z') \, \mathrm{d}\mu(z') = f(z) \qquad \forall f \in L^2(\mathcal{M}) . \tag{6.4}$$

We remark that this can always be achieved in terms of a Fourier representation in the following way. Consider a basis $\{\phi_j(z)\}$ of $L^2(\mathcal{M})$. Then for every $f \in L^2(\mathcal{M})$ we have the Fourier series

$$f(z) = \sum_j c_j \, \phi_j(z) \quad \text{with} \quad c_j = \int_M \overline{\phi_j(z)} \, f(z) \, \mathrm{d}\mu(z) . \tag{6.5}$$

Thus, comparing with (6.4), we find the Fourier representation of the δ-function,

$$\delta_{\mathcal{M}}(z - z') = \sum_j \phi_j(z) \, \overline{\phi_j(z')} . \tag{6.6}$$

Now the object of interest is the trace of U^t, defined by

$$\operatorname{Tr} U^t = \int_{\mathcal{M}} \delta_{\mathcal{M}}(\phi^t(z) - z) \, \mathrm{d}\mu(z) . \tag{6.7}$$

If the flow ϕ^t is ergodic we can replace a time average of $\operatorname{Tr} U^t$ by an $\mathcal{M}$-average, which can be easily calculated,

$$\lim_{T \to \infty} \frac{1}{T} \int_{0+}^{T} \operatorname{Tr} U^t \, \mathrm{d}t = \int_{\mathcal{M}} \int_{\mathcal{M}} \delta_{\mathcal{M}}(z' - z) \, \mathrm{d}\mu(z) \, \mathrm{d}\mu(z') = \int_{\mathcal{M}} \mathrm{d}\mu(z) = 1 . \tag{6.8}$$

The "+" in the lower bound of the t-integral indicates the exclusion of a small neighbourhood of $t = 0$.

On the other hand all contributions to the trace of U^t stem from times t which are equal to a period T_γ of a periodic orbit of ϕ^t. The integrated trace thus yields a periodic orbit sum of the form

$$\frac{1}{T}\int_{0+}^{T} \operatorname{Tr} U^t \, \mathrm{d}t = \frac{1}{T} \sum_{\gamma, T_\gamma \leq T} a_\gamma \tag{6.9}$$

with some amplitudes a_γ. Thus from the limit (6.8) we can deduce the classical sum rule

$$\sum_{\gamma, T_\gamma \leq T} a_\gamma \sim T\,, \quad T \to \infty\,. \tag{6.10}$$

6.2 Some Remarks on the Status of Sum Rules

The aim of this section is to briefly discuss some of the mathematical subtleties involved when one wants to derive classical sum rules in a more rigorous context. In particular we have to clarify the notion of trace used in (6.7) and in which sense convergence has to be understood in (6.8).

To this end consider again the time evolution operator U^t as defined in (6.1). Let us determine its adjoint,

$$\begin{aligned} \int_{\mathcal{M}} \overline{f(z)}\,\big(U^t g(z)\big)\,\mathrm{d}\mu(z) &= \int_{\mathcal{M}} \overline{f(z)}\, g(\phi^t(z))\,\mathrm{d}\mu(z) \\ &= \int_{\mathcal{M}} \overline{f(\phi^{-t}(z))}\, g(z)\,\mathrm{d}\mu(\phi^{-t}(z)) \\ &= \int_{\mathcal{M}} \left(U^{-t}\overline{f(z)}\right) g(z)\,\mathrm{d}\mu(z)\,, \end{aligned} \tag{6.11}$$

where we have used the invariance of μ. Since $U^{-t} = [U^t]^{-1}$ we see that the classical time evolution operator is unitary on $L^2(\mathcal{M})$.

We now employ the spectral decomposition of U^t, see e.g. [5, 6], keeping in mind that due to unitarity the spectrum is contained in the unit circle. Hence, we have

$$\begin{aligned} U^t &= \int_0^{2\pi} \mathrm{e}^{\mathrm{i}\omega t}\,\mathrm{d}E(\omega) \\ &= \sum_n \mathrm{e}^{\mathrm{i}\omega_n t} P_n + \int_0^{2\pi} \mathrm{e}^{\mathrm{i}\omega t}\,\mathrm{d}E_{\mathrm{ac}}(\omega) + \int_0^{2\pi} \mathrm{e}^{\mathrm{i}\omega t}\,\mathrm{d}E_{\mathrm{sc}}(\omega)\,, \end{aligned} \tag{6.12}$$

where $\exp(\mathrm{i}\omega_n)$ is an eigenvalue of U^1, P_n is a projector onto the corresponding eigenspace, and $E(\omega)$ is a (projector valued) spectral measure. The indices "ac" and "sc" refer to the absolutely continuous and singular continuous spectrum, respectively. The trace of U^t, in a distributional sense, can be taken as follows: Let $\{\varphi_n\}$ be an orthonormal set of eigenfunctions of U^t which we complement to an orthonormal basis of $L^2(\mathcal{M})$ by adding some $\{\psi_j\}$. Taking the trace in this basis yields,

$$\begin{aligned} \operatorname{Tr} U^t := \sum_n g_n \, \mathrm{e}^{\mathrm{i}\omega_n t} &+ \int_0^{2\pi} \mathrm{e}^{\mathrm{i}\omega t} \underbrace{\sum_j \langle \psi_j, \, \mathrm{d}E_{\mathrm{ac}}(\omega)\psi_j \rangle}_{=:\mathrm{d}\lambda_{\mathrm{ac}}(\omega)} \\ &+ \int_0^{2\pi} \mathrm{e}^{\mathrm{i}\omega t} \underbrace{\sum_j \langle \psi_j, \, \mathrm{d}E_{\mathrm{sc}}(\omega)\psi_j \rangle}_{=:\mathrm{d}\lambda_{\mathrm{sc}}(\omega)} , \end{aligned} \tag{6.13}$$

where g_n denotes the multiplicity of ω_n, and λ_{ac} and λ_{sc} are measures.

It can be shown that when averaging the trace as in (6.8) the first term, which derives from the pure point spectrum, yields 1 as in (6.8). This is due the fact that for an ergodic flow U^t has a non-degenerate eigenvalue 1, see e.g. [7, Chap. 2]. Determining the corresponding results for the two remaining terms can be difficult or even impossible. In general U^t need not be a trace-class operator and thus (6.13) may not make sense at all.

However recall that the quantity we are really interested in is the r.h.s of (6.7). As long as this is well-defined we can take (6.7) as a definition[1] for the expression $\operatorname{Tr} U^t$ on the l.h.s, which then is not necessarily related to the spectral trace of U^t, cf. [3] and also [8, Chap. VI, §2]. We may then proceed as sketched in Sect. 6.1.

Define

$$V^T : \; f(z) \;\mapsto\; \frac{1}{T} \int_{0+}^{T} f(\phi^t(z)) \, \mathrm{d}t \,, \tag{6.14}$$

i.e.

$$V^T = \frac{1}{T} \int_{0+}^{T} U^t \, \mathrm{d}t \tag{6.15}$$

with integral kernel

$$K_{V^T}(z, z') = \frac{1}{T} \int_{0+}^{T} K_{\mathrm{cl}}(z, z', t) \, \mathrm{d}t = \frac{1}{T} \int_{0+}^{T} \delta_{\mathcal{M}}(\phi^t(z) - z') \, \mathrm{d}t \,. \tag{6.16}$$

If ϕ^t is ergodic on $\mathcal{M}$ then

$$\lim_{T \to \infty} \left(V^T f \right)(z) = \int_{\mathcal{M}} f(z') \, \mathrm{d}\mu(z') \,, \tag{6.17}$$

$\forall \, f \in L^1(\mathcal{M})$ and μ-almost all $z \in \mathcal{M}$, i.e.

[1] I am grateful to Roman Schubert for pointing this out to me.

$$\lim_{T\to\infty} K_{V^T}(z, z') = 1 \tag{6.18}$$

in a weak sense. Now define the trace of V^T by

$$\operatorname{Tr} V^T := \int_{\mathcal{M}} K_{V^T}(z, z)\, \mathrm{d}\mu(z)\ . \tag{6.19}$$

From (6.18) we conclude that also

$$\lim_{T\to\infty} \operatorname{Tr} V^T = 1 \tag{6.20}$$

giving a clear meaning to (6.8) in terms of weak convergence. On the other hand

$$\operatorname{Tr} V^T = \int_{\mathcal{M}} \frac{1}{T} \int_{0+}^{T} \delta_{\mathcal{M}}(\phi^t(z) - z)\, \mathrm{d}t\, \mathrm{d}\mu(z)\ , \tag{6.21}$$

which yields a contribution whenever z lies on a periodic orbit. If these contributions are finite their explicit calculation will lead to classical sum rules, as we will show in the following.

6.3 Hannay-Ozorio de Almeida Sum Rules

We briefly discuss the derivation of the classical sum rules of Hannay and Ozorio de Almeida [1], see also [9], for strongly chaotic and integrable systems, respectively. The results of this section will later be used directly when we generalise the result to systems with spin.

6.3.1 Chaotic Systems

Let ϕ^t be a Hamiltonian flow on the phase space $\mathbb{R}^d \times \mathbb{R}^d$ which is ergodic on the energy shell Ω_E. In addition, all its periodic orbits shall be isolated and unstable. From (6.8) we know that

$$\lim_{T\to\infty} \frac{1}{T} \int_{0+}^{T} \operatorname{Tr} U^t\, \mathrm{d}t = 1\ , \tag{6.22}$$

where U^t acts on $L^2(\Omega_E)$. This can be compared with

$$\begin{aligned}
\lim_{T\to\infty} \frac{1}{T} \int_{0+}^{T} \operatorname{Tr} U^t\, \mathrm{d}t &= \lim_{T\to\infty} \frac{1}{T} \int_{0+}^{T} \int_{\Omega_E} \delta_{\Omega_E}(\phi^t(\boldsymbol{p}, \boldsymbol{x}) - (\boldsymbol{p}, \boldsymbol{x}))\, \mathrm{d}\mu_E(\boldsymbol{p}, \boldsymbol{x})\, \mathrm{d}t \\
&= \lim_{T\to\infty} \frac{1}{T} \sum_{\gamma, T_\gamma < T} \int_{T_\gamma -}^{T_\gamma +} \int_{\Omega_E} \delta_{\Omega_E}(\phi^t(\boldsymbol{p}, \boldsymbol{x}) - (\boldsymbol{p}, \boldsymbol{x}))\, \mathrm{d}\mu_E(\boldsymbol{p}, \boldsymbol{x})\, \mathrm{d}t\ ,
\end{aligned} \tag{6.23}$$

where in the last expression the sum extends over all periodic orbits γ with periods T_γ less than T and the t-integration is over a small interval around T_γ, which does not include periods of any periodic orbit other than γ, i.e. we assume that there are no accumulation points in the spectrum $\{T_\gamma\}$.

In the single integrals we now introduce new local coordinates with a (canonical) point transformation. We chose a coordinate u along the orbit (in configuration space) and $d-1$ coordinates v orthogonal to it. Their respective canonically conjugate momenta are denoted by p_u and p_v, i.e. points on the orbit (in phase space) are given by $(p_u, 0, u, 0)$. By solving $H(p_u, p_v, u, v) = E$ for p_u we can locally parameterise the energy shell Ω_E by (p_v, u, v), thus obtaining

$$\begin{aligned}
&\lim_{T\to\infty} \frac{1}{T} \int_{0+}^{T} \operatorname{Tr} U^t \, \mathrm{d}t \\
&= \lim_{T\to\infty} \frac{1}{T} \sum_{\gamma, T_\gamma < T} \int_{T_\gamma -}^{T_\gamma +} \int_{\Omega_E} \delta(u(t) - u(0))\, \delta\left(\begin{pmatrix} p_v(t) \\ v(t) \end{pmatrix} - \begin{pmatrix} p_v(0) \\ v(0) \end{pmatrix} \right) \\
&\qquad \times \mathrm{d}\mu_E(p_v(0), u(0), v(0))\, \mathrm{d}t \, .
\end{aligned} \tag{6.24}$$

For t close to T_γ,

$$\delta(u(t) - u(0)) = \frac{1}{|\dot{u}|}\, \delta(t - T_\gamma) \tag{6.25}$$

and

$$\begin{pmatrix} p_v(T_\gamma) \\ v(T_\gamma) \end{pmatrix} = \mathbb{M}_\gamma \begin{pmatrix} p_v(0) \\ v(0) \end{pmatrix} + \mathcal{O}\left(|p_v(0)|^2 + |v(0)|^2 \right) , \tag{6.26}$$

where $\mathbb{M}_\gamma$ is the monodromy matrix, the linearised Poincaré map on the Poincaré surface of section with $E, u = const.$ Performing the t-, p_v- and v-integration we are left with

$$\lim_{T\to\infty} \frac{1}{T} \int_{0+}^{T} \operatorname{Tr} U^t \, \mathrm{d}t = \lim_{T\to\infty} \frac{1}{T} \sum_{\gamma, T_\gamma < T} \int_\gamma \frac{\mathrm{d}u}{|\dot{u}|} \frac{1}{|\det(\mathbb{M}_\gamma - \mathbb{1}_{2d-2})|} , \tag{6.27}$$

where the remaining integration extends over all points on the orbit γ (in configuration space). If the periodic orbit is primitive we obtain its period as the result of the last integration (recall that $\dot{u}$ is the velocity). If it is the repetition of some shorter orbit, we only find the period of the latter, i.e. the primitive period $\mathbb{T}_\gamma^\#$ of the orbit γ. Comparing this result with (6.22) we obtain the Hannay-Ozorio de Almeida sum rule for strongly chaotic systems,

$$\sum_{\gamma, T_\gamma < T} \frac{T_\gamma^\#}{|\det(\mathbb{M}_\gamma - \mathbb{1}_{2d-2})|} \sim T \, , \quad T \to \infty \, . \tag{6.28}$$

Equivalently we have (formally by taking the derivative with respect to T)

$$\sum_{\gamma} \frac{T_{\gamma}^{\#}}{|\det(\mathbb{M}_{\gamma} - \mathbb{1}_{2d-2})|} \delta(T - T_{\gamma}) \sim 1 \,, \quad T \to \infty \,. \tag{6.29}$$

By multiplication with powers of T and subsequent integration we find the general rules

$$\sum_{\gamma} \frac{T_{\gamma}^{\#} T_{\gamma}^{n}}{|\det(\mathbb{M}_{\gamma} - \mathbb{1}_{2d-2})|} \delta(T - T_{\gamma}) \sim T^{n} \,, \quad T \to \infty \tag{6.30}$$

and

$$\sum_{\gamma, T_{\gamma} < T} \frac{T_{\gamma}^{\#} T_{\gamma}^{n}}{|\det(\mathbb{M}_{\gamma} - \mathbb{1}_{2d-2})|} \sim \frac{T^{n+1}}{n+1} \,, \quad T \to \infty \,. \tag{6.31}$$

6.3.2 Integrable Systems

At the other end of the scale are integrable systems which we will now discuss. For an integrable Hamiltonian flow ϕ_H^t we can locally introduce action and angle variables $(\boldsymbol{I}, \boldsymbol{\vartheta})$ such that the Hamiltonian depends only on the action variables $\boldsymbol{I}$. The dynamics then take place on a d-torus $\mathbb{T}^d = \mathbb{R}^d/(2\pi\mathbb{Z})^d$ which is parameterised by the angles $\boldsymbol{\vartheta}$ and the time evolution of the latter is determined by

$$\dot{\boldsymbol{\vartheta}} = \nabla_{\boldsymbol{I}} H(\boldsymbol{I}) =: \boldsymbol{\omega}(\boldsymbol{I}) \,. \tag{6.32}$$

These equations can be integrated trivially and we find

$$\phi_H^t(\boldsymbol{I}, \boldsymbol{\vartheta}) = (\boldsymbol{I}, \boldsymbol{\vartheta} + \boldsymbol{\omega} t) \,, \tag{6.33}$$

where here and in the following the variables $\boldsymbol{\vartheta}$ are defined modulo 2π. For the flow restricted to a particular torus (characterised by action variables $\boldsymbol{I}$) we also write

$$\phi_H^t(\boldsymbol{\vartheta}) = (\boldsymbol{\vartheta} + \boldsymbol{\omega} t) \,. \tag{6.34}$$

Again by (6.8) on an irrational and thus ergodic torus we have

$$\lim_{T \to \infty} \frac{1}{T} \int_{0+}^{T} \operatorname{Tr} U_{\boldsymbol{I}}^t \, \mathrm{d}t = 1 \,, \tag{6.35}$$

where $U_{\boldsymbol{I}}^t$ denotes the time evolution operator on $L^2(\mathbb{T}^d)$ and consequently the formal trace is defined by integration over $\mathbb{T}^d$, i.e.

$$\operatorname{Tr} U_{\boldsymbol{I}}^t = \int_{\mathbb{T}^d} \delta_{\mathbb{T}^d}(\phi_H^t(\boldsymbol{\vartheta}) - \boldsymbol{\vartheta}) \, \frac{\mathrm{d}^d \vartheta}{(2\pi)^d} = \int_{\mathbb{T}^d} \delta_{\mathbb{T}^d}(\boldsymbol{\omega}(\boldsymbol{I}) t) \, \frac{\mathrm{d}^d \vartheta}{(2\pi)^d} \,. \tag{6.36}$$

Note that in order to get the correct normalisation we need to define $\delta_{\mathbb{T}^d}(\boldsymbol{\vartheta}) := (2\pi)^d \, \delta^{(d)}(\boldsymbol{\vartheta})$. For a generic integrable system we expect that there are no systematic degeneracies in the frequencies $\boldsymbol{\omega}$. More precisely, we define a measure $\tilde{\mu}_E(\boldsymbol{I})$ on the energy shell in the set $\mathcal{I}$ of all action variables by

$$
\mathrm{d}\tilde{\mu}_E(\boldsymbol{I}) := \frac{(2\pi)^d}{|\Omega_E|}\delta(H(\boldsymbol{I})-E)\,\mathrm{d}^d I\,, \tag{6.37}
$$

which has the property that $\mathrm{d}\mu_E(\boldsymbol{I},\boldsymbol{\vartheta}) = \mathrm{d}\tilde{\mu}_E(\boldsymbol{I})\,\mathrm{d}^d\vartheta/(2\pi)^d$. We now assume that for $\tilde{\mu}_E$-almost all tori the frequencies $\boldsymbol{\omega}$ are linearly independent over $\mathbb{Q}$. Since the rational numbers are of measure zero in $\mathbb{R}$ we expect that this assumption should hold for systems that could be considered as generic in a reasonable sense. As we will see, the quantity to look at in oder to derive a sum rule is the $\tilde{\mu}_E$-average of (6.35),

$$
\begin{aligned}
\tilde{\mu}_E\left(\lim_{T\to\infty}\frac{1}{T}\int_{0+}^{T}\mathrm{Tr}\,U_{\boldsymbol{I}}^t\,\mathrm{d}t\right) &= \int_{H(\boldsymbol{I})=E}\lim_{T\to\infty}\frac{1}{T}\int_{0+}^{T}\mathrm{Tr}\,U_{\boldsymbol{I}}^t\,\mathrm{d}t\,\mathrm{d}\tilde{\mu}_E(\boldsymbol{I})\\
&= \lim_{T\to\infty}\frac{1}{T}\int_0^T\int_{\mathcal{I}}\int_{\mathbb{T}^d}\delta_{\mathbb{T}^d}(\boldsymbol{\omega}(\boldsymbol{I})t)\,\frac{\delta(H(\boldsymbol{I})-E)}{|\Omega_E|}\,\mathrm{d}^d\vartheta\,\mathrm{d}^d I\,.
\end{aligned} \tag{6.38}
$$

We perform the t- and $\boldsymbol{I}$-integrations which, due to the δ-functions, yield a sum over all rational tori with periods $T_{\boldsymbol{m}}$ and actions such that $H(\boldsymbol{I}) = E$. Note that

$$
\delta_{\mathbb{T}^d}(\boldsymbol{\omega}(\boldsymbol{I})t)\,\delta(H(\boldsymbol{I})-E) = \sum_{\boldsymbol{m}}\frac{(2\pi)^d}{|\det(\mathbb{D}_{\boldsymbol{m}})|}\delta(t-T_{\boldsymbol{m}})\delta^{(3)}(\boldsymbol{I}-\boldsymbol{I}_{\boldsymbol{m}}) \tag{6.39}
$$

where the required matrix of derivatives is given by, cf. eq. (3.115),

$$
\mathbb{D}_{\boldsymbol{m}} := \begin{pmatrix}\frac{\partial}{\partial t}(H(\boldsymbol{I})-E) & \nabla_{\boldsymbol{I}}(H(\boldsymbol{I})-E)\\ \frac{\partial}{\partial t}(\boldsymbol{\omega}(\boldsymbol{I})t) & \nabla_{\boldsymbol{I}}\otimes(\boldsymbol{\omega}(\boldsymbol{I})t)\end{pmatrix}\Bigg|_{\substack{\boldsymbol{I}=\boldsymbol{I}_{\boldsymbol{m}}\\ t=T_{\boldsymbol{m}}}} = \begin{pmatrix}0 & \boldsymbol{\omega}(\boldsymbol{I}_{\boldsymbol{m}})^T\\ \boldsymbol{\omega}(\boldsymbol{I}_{\boldsymbol{m}}) & \frac{\partial^2 H}{\partial I^2}T_{\boldsymbol{m}}\end{pmatrix}. \tag{6.40}
$$

Here $\partial^2 H/\partial I^2$ denotes the $d\times d$ matrix with entries

$$
\frac{\partial^2 H}{\partial I_j\partial I_k}(\boldsymbol{I}_{\boldsymbol{m}})\,,\quad j,k=1,\dots,d\,. \tag{6.41}
$$

Thus we find

$$
\tilde{\mu}_E\left(\lim_{T\to\infty}\frac{1}{T}\int_{0+}^{T}\mathrm{Tr}\,U_{\boldsymbol{I}}^t\,\mathrm{d}t\right) = \lim_{T\to\infty}\frac{1}{T}\sum_{\boldsymbol{m},T_{\boldsymbol{m}}<T}\int_{\mathbb{T}^d}\mathrm{d}^d\vartheta\,\frac{(2\pi)^d}{|\det\mathbb{D}_{\boldsymbol{m}}|\,|\Omega_E|}\,, \tag{6.42}
$$

and together with (6.35) this yields

$$
\begin{aligned}
&\lim_{T\to\infty}\frac{1}{T}\sum_{\boldsymbol{m},T_{\boldsymbol{m}}<T}\frac{(2\pi)^{2d}}{|\det\mathbb{D}_{\boldsymbol{m}}|\,|\Omega_E|} = 1\\
\Leftrightarrow\quad &\sum_{\boldsymbol{m},T_{\boldsymbol{m}}<T}\frac{(2\pi)^{2d}}{|\det\mathbb{D}_{\boldsymbol{m}}|}\sim T\,|\Omega_E|\,,\quad T\to\infty\,.
\end{aligned} \tag{6.43}
$$

By taking the derivative with respect to T we can also write the sum rule in the following way,

$$\sum_m \frac{(2\pi)^{2d}}{|\det \mathbb{D}_m|}\, \delta(T - T_m) \sim |\Omega_E| \,, \quad T \to \infty \,. \tag{6.44}$$

6.4 Classical Time Evolution Operators for Spinning Particles

Before we can derive the sum rules relevant for our further treatment of spectral two-point correlations we have to discuss how the method sketched above should be modified in order to correctly account for the spin dynamics. The time evolution operator for a classical particle with spin can be defined on $L^2(\mathbb{R}^{2d} \times S^2)$ by

$$U^t f(\boldsymbol{p}, \boldsymbol{x}, \boldsymbol{s}) = f\left(Y^t_{\mathrm{cl}}(\boldsymbol{p}, \boldsymbol{x}, \boldsymbol{s})\right) \,. \tag{6.45}$$

If we restrict the translational dynamics to the (compact) energy shell Ω_E, i.e. we consider U^t on $L^2(\Omega_E \times S^2)$, the skew product Y^t_{cl} preserves the product measure $\mu = \mu_E \times \mu_{S^2}$ of Liouville measure μ_E on the energy shell and the normalised surface element μ_{S^2} on the two-sphere. Thus, if, e.g, Y^t_{cl} is ergodic on $\Omega_E \times S^2$ all requirements of the previous sections are met. However, we have seen that functions on S^2 which are classical observables corresponding to quantum observables for a spin s are restricted to certain subspaces, see Sect. 4.2. Therefore, we should consider U^t not as an operator defined on $L^2(\Omega_E \times S^2)$ but rather on the subspaces $L^2(\Omega_E) \otimes \mathcal{H}_s$ with

$$\mathcal{H}_s := \langle \{Y_{lm}(\boldsymbol{s})\}_{l \leq 2s} \rangle \,. \tag{6.46}$$

This implies a modification when defining the integral kernel of U^t. Recall that to this end we need a δ-function on S^2. Using the spherical harmonics Y_{lm} this can be achieved explicitly by

$$\delta_{S^2}(\boldsymbol{s} - \boldsymbol{s}') := 4\pi \sum_{l=0}^{\infty} \sum_{m=-l}^{l} Y_{lm}(\boldsymbol{s})\, \overline{Y_{lm}(\boldsymbol{s}')} \,. \tag{6.47}$$

However, since higher harmonics ($l > 2s$) are not in our Hilbert space we also have to truncate this expansion, i.e. we define the reproducing kernel

$$\delta_s(\boldsymbol{s}, \boldsymbol{s}') := 4\pi \sum_{l=0}^{2s} \sum_{m=-l}^{l} Y_{lm}(\boldsymbol{s})\, \overline{Y_{lm}(\boldsymbol{s}')} \tag{6.48}$$

(note that we have not added an additional index "S^2", since this should be implied from the context). Clearly, when folded with a test function of the respective space, we still have the desired property

$$\int_{S^2} f(s')\,\delta_s(s,s')\,\mathrm{d}\mu_{S^2}(s') = f(s) \quad \forall\, f\in\mathcal{H}_s\,. \tag{6.49}$$

Later we will briefly discuss the result one obtains using δ_{S^2} instead of δ_s.

In order to demonstrate how to work explicitly with these δ-functions and for later reference we calculate the integral

$$I_s(\boldsymbol{n},\alpha) := \int_{S^2} \delta_s(\boldsymbol{s}, R(\boldsymbol{n},\alpha)\boldsymbol{s})\,\mathrm{d}\mu_{S^2}(\boldsymbol{s}) \tag{6.50}$$

where $R(\boldsymbol{n},\alpha)\in\mathrm{SO}(3)$ is a rotation about an axis $\boldsymbol{n}$ by an angle α. Choosing the $\boldsymbol{s}$-coordinate system such that the s_z-axis is parallel to $\boldsymbol{n}$ and then introducing spherical coordinates (θ,ϕ) as in (4.15) we find

$$R(\boldsymbol{n},\alpha)\,\boldsymbol{s}(\theta,\phi) = \boldsymbol{s}(\theta,\phi+\alpha)\,. \tag{6.51}$$

Hence, the integral is given by

$$\begin{aligned} I_s(\boldsymbol{n},\alpha) &= \int_0^{2\pi}\int_0^{\pi}\sum_{l=0}^{2s}\sum_{m=-l}^{l} Y_{lm}(\theta,\phi)\,\overline{Y_{lm}(\theta,\phi+\alpha)}\,\sin\theta\,\mathrm{d}\theta\,\mathrm{d}\phi \\ &= \sum_{l=0}^{2s}\sum_{m=-l}^{l}\int_0^{2\pi} \mathrm{e}^{\mathrm{i}m\phi}\,\mathrm{e}^{-\mathrm{i}m(\phi+\alpha)}\,\frac{\mathrm{d}\phi}{2\pi} = \sum_{l=0}^{2s}\sum_{m=-l}^{l}\mathrm{e}^{-\mathrm{i}m\alpha}\,, \end{aligned} \tag{6.52}$$

and, as one might have expected, does not depend on $\boldsymbol{n}$ but only on the angle α. Thus, in the following we suppress the argument $\boldsymbol{n}$ by writing $I_s(\alpha)\equiv I_s(\boldsymbol{n},\alpha)$. The double sum can be simplified further to

$$I_s(\alpha) = \sum_{m=-2s}^{2s}\sum_{l=|m|}^{2s}\mathrm{e}^{\mathrm{i}m\alpha} = \sum_{m=-2s}^{2s}(2s+1-|m|)\,\mathrm{e}^{\mathrm{i}m\alpha}\,. \tag{6.53}$$

The corresponding integral with δ_s replaced by δ_{S^2} can be calculated as follows,

$$\begin{aligned} I_\infty(\alpha) &:= \int_{S^2}\delta_{S^2}(\boldsymbol{s}-R(\boldsymbol{n},\alpha)\boldsymbol{s})\,\mathrm{d}\mu_{S^2}(\boldsymbol{s}) \\ &= \int_{\mathbb{R}^3}\delta(|\boldsymbol{s}|-1)\,\delta_{S^2}(\boldsymbol{s}-R(\boldsymbol{n},\alpha)\boldsymbol{s})\,\mathrm{d}^3s \\ &= \int_{\mathbb{R}^3}\delta^{(3)}\left((\mathbb{1}_3-R(\boldsymbol{n},\alpha))\boldsymbol{s}\right)\,\mathrm{d}^3s = \left|\det(\mathbb{1}_3-R(\boldsymbol{n},\alpha)\right|^{-1/2} \\ &= \left|(1-\mathrm{e}^{\mathrm{i}\alpha})(1-\mathrm{e}^{-\mathrm{i}\alpha})\right|^{-1/2} = \frac{1}{4\sin^2(\alpha/2)}\,. \end{aligned} \tag{6.54}$$

6.5 Spin in Classical Sum Rules

In order to derive classical sum rules for the skew product flow Y^t_{cl} we combine the results from Sects. 6.3 and 6.4.

6.5.1 Chaotic Systems

Let us first consider the case when Y_{cl}^t is ergodic on $\Omega_E \times S^2$. Then (6.8) still guarantees that

$$\lim_{T\to\infty} \frac{1}{T} \int_{0+}^{T} \operatorname{Tr} U^t \, \mathrm{d}t = 1 \, , \tag{6.55}$$

where now U^t acts on $L^2(\Omega_E) \otimes \mathcal{H}_s$ and Tr denotes the respective formal trace. Explicitly we have

$$\begin{aligned} \lim_{T\to\infty} \frac{1}{T} \int_{0+}^{T} \operatorname{Tr} U^t \, \mathrm{d}t = \lim_{T\to\infty} \frac{1}{T} \int_{0+}^{T} \int_{\Omega_E} \int_{S^2} \delta_{\Omega_E} \left(\phi^t(\boldsymbol{p},\boldsymbol{x}) - (\boldsymbol{p},\boldsymbol{x}) \right) \\ \times \, \delta_s(R(\boldsymbol{p},\boldsymbol{x},t)\boldsymbol{s},\boldsymbol{s}) \, \mathrm{d}\mu_{S^2}(\boldsymbol{s}) \, \mathrm{d}\mu_E(\boldsymbol{p},\boldsymbol{x}) \, \mathrm{d}t \, . \end{aligned} \tag{6.56}$$

The $\boldsymbol{s}$-integral was calculated in the previous section, (6.50) – (6.53), so we are left with the evaluation of

$$\lim_{T\to\infty} \frac{1}{T} \int_{0+}^{T} \int_{\Omega_E} \delta_{\Omega_E} \left(\phi^t(\boldsymbol{p},\boldsymbol{x}) - (\boldsymbol{p},\boldsymbol{x}) \right) \, I_s(\alpha(\boldsymbol{p},\boldsymbol{x},t)) \, \mathrm{d}\mu_E(\boldsymbol{p},\boldsymbol{x}) \, \mathrm{d}t \, , \tag{6.57}$$

where $\alpha(\boldsymbol{p},\boldsymbol{x},t)$ is the angle by which a spin vector has been rotated after precessing up to time t along the trajectory starting at $(\boldsymbol{p},\boldsymbol{x})$. In Sect. 6.3 it was shown that all contributions to the remaining integrals derive from periodic orbits γ of the the flow ϕ_H^t, and in Sect. 3.4 we have seen that the angle $\alpha(\boldsymbol{p},\boldsymbol{x},T_\gamma)$ for $(\boldsymbol{p},\boldsymbol{x}) \in \gamma$ does not depend on the point $(\boldsymbol{p},\boldsymbol{x})$ on the orbit, i.e. we may define

$$\alpha_\gamma := \alpha(\boldsymbol{p},\boldsymbol{x},T_\gamma) \, , \quad (\boldsymbol{p},\boldsymbol{x}) \in \gamma \, . \tag{6.58}$$

Combing this knowledge with the result (6.27) we can state the sum rule in the present case,

$$\sum_{\gamma, T_\gamma < T} \frac{T_\gamma^\# \, I_s(\alpha_\gamma)}{|\det(\mathbb{M}_\gamma - \mathbb{1}_{2d-2})|} \sim T \, , \quad T \to \infty \, , \tag{6.59}$$

where all quantities have the same meaning as in (6.27). In all previous cases we have found that the weight factor in the classical sum is (besides powers of $T_\gamma^\#$) the square of the corresponding term in the semiclasscial trace formula. Therefore, we will investigate the relation between $I_s(\alpha)$ and $\chi_s(\alpha)$ more closely. Recall that the character of the $(2s+1)$-dimensional unitary irreducible representation of SU(2) parametrised by the rotation angle α is given by, cf. Appendix C,

$$\chi_s(\alpha) = \sum_{m=-s}^{s} \mathrm{e}^{\mathrm{i}m\alpha} \, , \tag{6.60}$$

where the summation index m is always raised by steps of one, for both integer and half-integer s. On the other hand we have found (6.53)

$$I_s(\alpha) = \sum_{m=-2s}^{2s} (2s+1-|m|)\,\mathrm{e}^{\mathrm{i}m\alpha} . \tag{6.61}$$

Now,

$$\chi_s^2(\alpha) = \sum_{m=-s}^{s} \sum_{m'=-s}^{s} \mathrm{e}^{\mathrm{i}(m+m')\alpha} , \tag{6.62}$$

and introducing a new index $M := m + m'$, which runs from $-2s$ to $2s$, the exponential is of the same form as in (6.60); moreover, there are exactly $2s+1-|M|$ possibilities for obtaining a given M from m and m' both ranging from $-s$ to s. Thus we find

$$\chi_s^2(\alpha) = I_s(\alpha) \tag{6.63}$$

as expected. Finally we may write the distributional version of the classical sum rule for ergodic skew products Y_{cl}^t as follows,

$$\sum_\gamma \frac{T_\gamma^\# \,\chi_s^2(\alpha)}{|\det(\mathbb{M}_\gamma - \mathbb{1}_{2d-2})|}\,\delta(T-T_\gamma) \sim 1 , \quad T\to\infty . \tag{6.64}$$

6.5.2 Integrable Systems

We can now turn our attention to the integrable case as discussed in 4.4.2. Recall that the total phase space $\mathbb{R}^d \times \mathbb{R}^d \times S^2$ foliates into invariant manifolds which are bundles $\mathcal{T}_\theta \xrightarrow{\pi} \mathbb{T}^d$ over Liouville–Arnold tori with typical fibre S^1. Furthermore, an embedding of the fibres in S^2 was given in terms of parallels of latitude. If the respective torus is ergodic (i.e. $\{\omega_j/\pi\}$ rationally independent) and in addition, for the rotation angles α_j corresponding to the basis cycles, we also have that the $\{\alpha_j/\pi\}$ are rationally independent, then the skew product flow restricted to the corresponding latitude bundle is also ergodic.

In a slight variation of our previous strategy we will not define a time evolution operator on $\mathcal{T}_\theta$ but investigate the classical time evolution on $\mathbb{T}^2 \times S^2$. In that way we treat the dynamics on all $\mathcal{T}_\theta$, $0 \le \theta \le \pi$, given by the same rotations $R(\boldsymbol{n}(\boldsymbol{\vartheta},\alpha))$, simultaneously. Consider

$$\lim_{T\to\infty} \frac{1}{T} \int_{0+}^{T} \mathrm{Tr}\, U^t \,\mathrm{d}t \tag{6.65}$$

where U^t propagates observables $f \in L^2(\mathbb{T}^d) \otimes \mathcal{H}_s$ and Tr denotes the respective formal trace, i.e.

$$
\begin{aligned}
&\lim_{T\to\infty}\frac{1}{T}\int_{0+}^{T}\operatorname{Tr}U^t\,\mathrm{d}t \\
&=\lim_{T\to\infty}\frac{1}{T}\int_{0+}^{T}\int_{\mathbb{T}^d}\delta_{\mathbb{T}^d}(\phi_H^t(\boldsymbol{\vartheta})-\boldsymbol{\vartheta})\int_{S^2}\delta_s(R(\boldsymbol{I},\boldsymbol{\vartheta},t)\boldsymbol{s},\boldsymbol{s})\,\mathrm{d}\mu_{S^2}(\boldsymbol{s})\,\mathrm{d}\mu_{\mathbb{T}^d}(\boldsymbol{\vartheta})\,\mathrm{d}t\,.
\end{aligned}
\tag{6.66}
$$

In the $\boldsymbol{s}$-integral we introduce spherical coordinates (θ,ϕ) such that the s_z-axis is parallel to $\boldsymbol{n}(\boldsymbol{I},\boldsymbol{\vartheta})$, i.e.

$$
\int_{S^2}\delta_s(R(\boldsymbol{I},\boldsymbol{\vartheta},t)\boldsymbol{s},\boldsymbol{s})\,\mathrm{d}\mu_{S^2}(\boldsymbol{s})=\int_{S^2}\delta_s(\boldsymbol{s}(\theta,\phi+\alpha(\boldsymbol{I},\boldsymbol{\vartheta},t)),\boldsymbol{s}(\theta,\phi))\mathrm{d}\mu(\boldsymbol{s})\,.
\tag{6.67}
$$

We have not indicated that the choice of variables in the $\boldsymbol{s}$-integral depends on the point $\boldsymbol{\vartheta}$ on $\mathbb{T}^d$, so we have to keep in mind that we must perform the $\boldsymbol{s}$-integration before the $\boldsymbol{\vartheta}$-integration. Ergodicity of Y_{cl}^t on $\mathcal{T}_\theta\xrightarrow{\pi}\mathbb{T}^d$ means

$$
\begin{aligned}
&\lim_{T\to\infty}\frac{1}{T}\int_{0+}^{T}\operatorname{Tr}U^t\,\mathrm{d}t \\
&=\int_{\mathbb{T}^d}\int_{\mathbb{T}^d}\delta_{\mathbb{T}^d}(\boldsymbol{\vartheta}'-\boldsymbol{\vartheta})\int_0^{2\pi}\int_0^{\pi}\int_0^{2\pi}\delta_s(\boldsymbol{s}(\theta,\phi'),\boldsymbol{s}(\theta,\phi)) \\
&\qquad\frac{\mathrm{d}\phi}{2\pi}\,\frac{\sin\theta\,\mathrm{d}\theta}{2}\,\frac{\mathrm{d}\phi'}{2\pi}\,\mathrm{d}\mu_{\mathbb{T}^d}(\boldsymbol{\vartheta})\,\mathrm{d}\mu_{\mathbb{T}^d}(\boldsymbol{\vartheta}')\,.
\end{aligned}
\tag{6.68}
$$

With the definition (6.48) the integrals over θ,ϕ and ϕ' can be carried out explicitly, yielding

$$
\begin{aligned}
&\int_0^{2\pi}\int_0^{\pi}\int_0^{2\pi}\delta_s(\boldsymbol{s}(\theta,\phi'),\boldsymbol{s}(\theta,\phi))\frac{\mathrm{d}\phi}{2\pi}\,\frac{\sin\theta\,\mathrm{d}\theta}{2}\,\frac{\mathrm{d}\phi'}{2\pi} \\
&=\int_0^{2\pi}\int_0^{2\pi}\sum_{l=0}^{2s}\sum_{m=-l}^{l}\mathrm{e}^{\mathrm{i}m(\phi-\phi')}\,\frac{\mathrm{d}\phi}{2\pi}\,\frac{\mathrm{d}\phi'}{2\pi}=2s+1\,.
\end{aligned}
\tag{6.69}
$$

Thus, for the whole expression we also find

$$
\lim_{T\to\infty}\frac{1}{T}\int_{0+}^{T}\operatorname{Tr}U^t\,\mathrm{d}t=2s+1\,.
\tag{6.70}
$$

On the other hand we must to determine the contributions of periodic orbits, i.e. of rational tori. To this end consider the $\tilde{\mu}_E$ average of (6.66) as in (6.38). If on a given torus the fundamental frequencies are pairwise rationally dependent, all orbits on the torus are periodic with period $T_{\boldsymbol{m}}$ and the rotation angle $\alpha(\boldsymbol{I},\boldsymbol{\vartheta},T_{\boldsymbol{m}})$ is the same at each point $\boldsymbol{\vartheta}$ of the torus, cf. the discussion in Sects. 3.5.3 and 5.3. Therefore, we may define

$$
\alpha_{\boldsymbol{m}}:=\alpha(\boldsymbol{I}_{\boldsymbol{m}},\boldsymbol{\vartheta},T_{\boldsymbol{m}})\,.
\tag{6.71}
$$

Now we can split the integrals in (6.66) as in (6.38), and the spin contribution is given by $I_s(\alpha_{\boldsymbol{m}}) = \chi_s^2(\alpha_{\boldsymbol{m}})$. Finally, we obtain the sum rule

$$\sum_{\boldsymbol{m},T_{\boldsymbol{m}}<T} \frac{(2\pi)^{2d}\chi_s^2(\alpha_{\boldsymbol{m}})}{|\det \mathbb{D}_{\boldsymbol{m}}|} \sim (2s+1)\,|\Omega_E|\,T\,, \quad T\to\infty\,, \tag{6.72}$$

or, by taking the derivative with respect to T,

$$\sum_{\boldsymbol{m}} \frac{(2\pi)^{2d}\chi_s^2(\alpha_{\boldsymbol{m}})}{|\det \mathbb{D}_{\boldsymbol{m}}|}\,\delta(T-T_{\boldsymbol{m}}) \sim (2s+1)\,|\Omega_E|\,, \quad T\to\infty\,. \tag{6.73}$$

6.5.3 Partially Integrable Systems

As a last example for classical sum rules with spin we want to discuss an intermediate case. Consider the situation where ϕ_H^t is an integrable Hamiltonian flow but the skew product Y_{cl}^t is not integrable in the sense of Definition 1. Instead the spin dynamics shall be such that for $\tilde{\mu}_E$-almost all tori $\mathbb{T}^d$ the skew product Y_{cl}^t is ergodic on $\mathbb{T}^d \times S^2$ with respect to $\mu_{\mathbb{T}^d} \times \mu_{S^2}$. By our previous general consideration (6.8) we find

$$\lim_{T\to\infty} \frac{1}{T}\int_{0+}^{T} \mathrm{Tr}\, U^t\,\mathrm{d}t = 1 \tag{6.74}$$

(where the trace is defined as in Section 6.5.2) for $\tilde{\mu}_E$-almost all tori characterised by actions $\boldsymbol{I}$. We also find that

$$\begin{aligned} &\tilde{\mu}_E\left(\lim_{T\to\infty}\frac{1}{T}\int_0^T \mathrm{Tr}\,U^t\,\mathrm{d}t\right) \\ &= \int_{H(\boldsymbol{I})=E} \lim_{T\to\infty}\frac{1}{T}\int_0^T\int_{\mathbb{T}^d}\int_{S^2} \delta_{\mathbb{T}^d}(\boldsymbol{\omega}(\boldsymbol{I})\,t)\,\delta_s(R(\boldsymbol{I},\boldsymbol{\vartheta},t)\boldsymbol{s},\boldsymbol{s}) \\ &\qquad\qquad \times\,\mathrm{d}\mu_{S^2}(\boldsymbol{s})\,\mathrm{d}\mu_{\mathbb{T}^d}(\boldsymbol{\vartheta})\,\mathrm{d}t\,\mathrm{d}\tilde{\mu}_E(\boldsymbol{I})\,. \end{aligned} \tag{6.75}$$

For a fixed $(\boldsymbol{I},\boldsymbol{\vartheta})$ we can still introduce coordinates on S^2 such that the rotation axis of $R(\boldsymbol{I},\boldsymbol{\vartheta},t)$ coincides with the s_z-direction, yielding

$$\int_{S^2}\delta_s(R(\boldsymbol{I},\boldsymbol{\vartheta},t)\boldsymbol{s},\boldsymbol{s})\,\mathrm{d}\mu_{S^2}(\boldsymbol{s}) = I_s(\alpha(\boldsymbol{I},\boldsymbol{\vartheta},t)) = \chi_s^2(\alpha(\boldsymbol{I},\boldsymbol{\vartheta},t))\,. \tag{6.76}$$

But now, even for $t=T_{\boldsymbol{m}}$, in general $\alpha(\boldsymbol{I}_{\boldsymbol{m}},\boldsymbol{\vartheta},T_{\boldsymbol{m}})$ is not constant on the corresponding torus. Performing the remaining integrals as in the previous cases, cf. (6.38) and thereafter, we obtain a sum rule in which the spin contribution is averaged over the respective rational torus,

$$\sum_{\boldsymbol{m},T_{\boldsymbol{m}}<T} \frac{(2\pi)^{2d}\,\overline{\chi_s^2}^{\boldsymbol{m}}}{|\det \mathbb{D}_{\boldsymbol{m}}|} \sim |\Omega_E|T\,, \quad T\to\infty\,, \tag{6.77}$$

with

$$\overline{\chi_s^2}^{\boldsymbol{m}} = \frac{1}{(2\pi)^d} \int_{\mathbb{T}^d} \chi_s^2(\alpha(\boldsymbol{I}_{\boldsymbol{m}}, \boldsymbol{\vartheta}, T_{\boldsymbol{m}}))\, \mathrm{d}\mu_{\mathbb{T}^d}(\boldsymbol{\vartheta})\,. \tag{6.78}$$

Again by taking the derivative with respect to T the sum rule can be written as follows,

$$\sum_{\boldsymbol{m}} \frac{(2\pi)^{2d}\, \overline{\chi_s^2}^{\boldsymbol{m}}}{|\det \mathbb{D}_{\boldsymbol{m}}|}\, \delta(T - T_{\boldsymbol{m}}) \sim |\Omega_E|\,, \quad T \to \infty\,. \tag{6.79}$$

References

1. J.H. Hannay, A.M. Ozorio de Almeida: J. Phys. A **17**, 3429–3440 (1984)
2. W. Parry, M. Pollicott: Astérisque **187-8**, 1–268 (1990)
3. V. Guillemin: Duke Math. J. **44**, 485–517 (1977)
4. P. Cvitanović, B. Eckhardt: J. Phys. A **24**, L237–L241 (1991)
5. M. Reed, B. Simon: *Methods of Modern Mathematical Physics I, Functional Analysis* (Academic Press, San Diego, 1980)
6. W. Rudin: *Functional Analysis*, 2nd edn. (McGraw-Hill, New York, 1991)
7. Ya.G. Sinai (Ed.): *Dynamical Systems II, Ergodic Theory with Applications to Dynamical Systems and Statistical Mechanics*, Vol. 2 of Encyclopaedia of Mathematical Sciences (Springer-Verlag, Berlin, 1989)
8. V. Guillemin, S. Sternberg: *Geometric asymptotics* (American Mathematical Society, Providence, R.I., 1977), mathematical Surveys, No. 14
9. A.M. Ozorio de Almeida: *Hamiltonian Systems: Chaos and Quantization* (Cambridge University Press, Cambridge, 1988)

7 Spectral Statistics and Spin

In this chapter we apply the results collected in the previous chapters to an analysis of spectral statistics.

The research field of quantum chaos consistes of the search for fingerprints of classical chaos in quantum mechanics. In classical mechanics chaoticity is characterised by the long time behaviour of a system, e.g. by the exponentially growing separation of trajectories started with neighbouring initial conditions. In quantum mechanics the long time behaviour of a bound system (with a purely discrete spectrum) is always quasi-periodic – a property which in classical mechanics is characteristic of integrable systems. Therefore, the connection between classically chaotic dynamics and spectral statitics has to be established in a more indirect way.

In 1979 McDonald and Kaufman found [1] that the statistical distribution of energy eigenvalues depends sensibly on whether the corresponding classical system shows integrable or chaotic dynamics. Their observation was subsequently verified for various systems, see e.g. [2–5]. The hypothesis, that chaotic classical dynamics give rise to spectral statistics which can be described by random matrix theory nowadays goes under the name Bohigas–Giannoni–Schmit (BGS) conjecture after [3]. The respective statement for integrable systems, which asserts that their corresponding quantum spectral statistics can be described by a Poisson process, is due Berry and Tabor [6].

In random matrix theory, see e.g. [7, 8], one investigates the statistical distribution of eigenvalues in ensembles of matrices with random entries. The only restrictions on the matrix elements are posed by certain symmetry requirements. For instance one defines the Gaussian unitary, orthogonal and symplectic ensembles (GUE, GOE and GSE) consisting of either hermitian, real symmetric or quaternion real matrices, respectively, whose entries are independent Gaussian random variables. Random matrix theory was developed in order to describe spectra in nuclear physics, where the precise form of the complicated multi-particle interactions was unknown. Therefore, the Hamiltonian was replaced by a random matrix, and the symmetry requirements correspond to time reversal invariant situations with integer (GOE) or half-integer (GSE) total angular momentum and to situations in which time reversal invariance is broken (GUE); see also the discussion in the following section.

The aforementioned application of random matrix theory in the field of quantum chaos was guided by the same principles, whereby, roughly speaking, classically chaotic dynamics replaced the role of complicated multi-particle interactions. Reviews of quantum chaos in general and of semiclassics and spectral statistics in particular can be found in [4, 5, 9–15].

A connection between the quantum spectral statistics and the characteristics of the corresponding classical system was established by Hannay and Ozorio de Almeida [16] and subsequently by Berry [17]. These authors used the Gutzwiller trace formula in order to obtain a semiclassical expression for the spectral two-point correlation function (or its Fourier transform, the spectral form factor). They then invoked the so-called diagonal approximation and with the help of classical sum rules (see [16] and Chap. 6) were able to connect the different spectral statistics with integrable and chaotic classical dynamics, respectively.

We should remark that there are also examples for which the BGS-conjecture fails to predict the correct statistics [18–20]. These examples are known under the notion of arithmetical chaos. The discrepancy can also be understood on the basis of trace formulae [21–23], which in these cases are exact. In general one expects the universality conjectures formulated in [6] and [3] to hold only for generic systems. However, a precise measure theoretical or topological definition of the meaning of genericity in this context is still missing.

We also want to mention two different approaches which extract information on quantum spectral statistics going beyond the results of [16, 17]. The first one is a bootstrap technique due to Bogomolny and Keating [24] which relates some of the terms neglected by the diagonal approximation to the diagonal terms. The second approach is due to Richter and Sieber [25] and Sieber [26, 27]. It exploits information on correlations of certain (figure-eight shaped) pairs of periodic orbits. Preliminary results indicate that with the methods presented in this work the approach of Richter and Sieber carries over to the case of systems with spin [28].

In Sect. 7.3 we describe a semiclassical theory for the spectral form factor based on regularised trace formulae for the Pauli and the Dirac equation, cf. [29]. We discuss the so-called diagonal approximation (Sect. 7.3.1), which was first used in [16], and its limitations. The results which we obtain in Sects. 7.3.2 and 7.3.3 for classically chaotic and integrable systems, respectively, generalise the known results [16, 17] for systems with spin 0, to arbitrary spin in the integrable case and to arbitrary integer spin in the chaotic case. For chaotic systems with half-integer spin one expects statistics according to the GSE which is also supported by the respective semiclassical results, see Sect. 7.3.2 and [29]. In addition we discuss a partially integrable situation, in which the translational dynamics are integrable but not the skew product of translational and spin dynamics (Sect. 7.3.4). Our semiclassical theory predicts intermediate statistics for this case which we illustrate with an example in Sect. 7.4. In Sect. 7.5 we briefly discuss semiclassical theories for other statistical functions apart from the spectral form factor. There we extend previously known results for spin 0, that relate to the GOE and GUE, respectively, to arbitrary spin for which also the GSE appears. Thus, we mostly focus on results relating to the latter.

7.1 Symmetries and Unfolding

When we want to compare the spectral correlations of a particular quantum system with universal distributions we have to take care of two aspects first, namely desymmetrisation and unfolding, in order to make spectra comparable.

We say that the system under consideration is invariant under a particular (geometrical) symmetry, if the Hamiltonian commutes with a unitary representation of the respective (finite) symmetry group. If there are subspaces invariant under the action of the symmetry group, the spectrum of the Hamiltonian should be considered in a single subspace only. Partial spectra corresponding to different subspaces must be treated as independent, since there is no interaction between states corresponding to different subspaces (the Hamiltonian is block diagonal). E.g., if the system under consideration is invariant under parity, $[\hat{P},\hat{H}] = 0$, $\hat{P}\psi(\boldsymbol{x}) = \psi(-\boldsymbol{x})$, one should discuss spectral statistics only for the energy eigenvalues corresponding to states with either positive or negative parity. This process is known as desymmetrisation. In the following we will always assume that (chaotic) systems have been desymmetrised.

Notice that this discussion does not make sense for integrable systems. If we have conserved quantities, hermitian operators commuting with the Hamiltonian, we can consider them as (the derived representation of) generators of a continuous group. Desymmetrisation with respect to this group would imply that we discuss spectral statistics only for a subspectrum corresponding to states with given eigenvalues for all conserved quantities but the Hamiltonian. Such a spectrum is equivalent to the spectrum of a one-dimensional system which does not show the generic features we are interested in. Therefore, one should not attempt a complete desymmetrysation of integrable systems. Instead desymmetrisation in this case is reduced to the task of finding a subspace on which the Hamiltonian $\hat{H}$ has no systematic degeneracies (cf. also the discussion of Kramers' degeneracy below). For the resulting system, in which there are still symmetries present, we can then expect the same statistics as for a Poisson process.

Another situation which has to be treated differently is that of antiunitary symmetries, like time reversal. The time reversal operator is given by

$$\hat{T} = \mathrm{e}^{(\mathrm{i}/\hbar)\pi\hat{s}_y}\,\hat{K}\,,^{1} \tag{7.1}$$

[1] The reasoning directly carries over to the case of the Dirac equation, if for the spin operator we choose $\hat{s} = \frac{\hbar}{2}\left(\begin{smallmatrix}\sigma & 0\\ 0 & \sigma\end{smallmatrix}\right)$.

where $\hat{K}$ denotes the operator of complex conjugation. Of course, this definition depends on the representation we are working in. Eq. (7.1) applies if we use the position representation ($\hat{\boldsymbol{x}} = \boldsymbol{x}$, $\hat{\boldsymbol{p}} = (\hbar/\mathrm{i})\nabla$) and if we choose the spin operators such that $\hat{s}_x$ and $\hat{s}_z$ are real and $\hat{s}_y$ is purely imaginary. This choice is consistent with using $\hat{\boldsymbol{s}} = (\hbar/2)\boldsymbol{\sigma}$ for a spin-1/2 system with the standard definition of the Pauli matrices $\boldsymbol{\sigma}$, see Appendix C, in which case (7.1) reduces to $\hat{T} = \mathrm{i}\sigma_y \hat{K}$. One easily verifies that

$$\hat{T}^2 = (-1)^{2s} . \tag{7.2}$$

By direct computation one also checks that

$$\begin{aligned} \hat{T}\,\hat{\boldsymbol{x}}\,\hat{T}^{-1} &= \hat{\boldsymbol{x}} , \\ \hat{T}\,\hat{\boldsymbol{p}}\,\hat{T}^{-1} &= -\hat{\boldsymbol{p}} , \\ \hat{T}\,\hat{\boldsymbol{s}}\,\hat{T}^{-1} &= -\hat{\boldsymbol{s}} . \end{aligned} \tag{7.3}$$

This is why $\hat{T}$ is called the time reversal operator: Time reversal inverts momentum but leaves the position invariant. Thus it also inverts angular momentum, and we have defined $\hat{T}$ such that, in analogy, it also inverts spin. We say that the system defined by a quantum Hamiltonian $\hat{H}$ is time reversal invariant if $[\hat{H}, \hat{T}] = 0$. Unlike the situation in the case of unitary symmetries, in general there are in no invariant subspaces of our total Hilbert space $\mathcal{H}$. In order to see this recall that $\hat{T}$ being anti-linear means that $\hat{T}(\mu\psi) = \overline{\mu}\psi$, $\mu \in \mathbb{C}$, $\psi \in \mathcal{H}$. Therefore, if the pair (ψ, λ) fulfills an "eigenvalue equation" of $\hat{T}$, i.e. $\hat{T}\psi = \lambda\psi$, it follows that $\mathrm{i}\psi$ is an "eigenfunction" with the eigenvalue $-\lambda$; even worse, for $\mu \in \mathbb{C}$ neither real nor purely imaginary, $\mu\psi$ does not fulfill an "eigenvalue equation" for $\hat{T}$ at all. Thus, we conclude that we can not desymmetrise with respect to anti-unitary symmetry operations. Instead one has to keep in mind that the system either is time reversal invariant or not, which will give rise to different spectral statistics. We have to distinguish three symmetry classes:

1. $[\hat{T}, \hat{H}] \neq 0$, i.e. the system is not invariant under time reversal.
2. $[\hat{T}, \hat{H}] = 0$, $\hat{T}^2 = 1$, i.e. the system is invariant under time reversal and has integer spin s.
3. $[\hat{T}, \hat{H}] = 0$, $\hat{T}^2 = -1$, i.e. the system is invariant under time reversal and has half-integer spin s.

The last case is particularly interesting, because it does not occur for spinless systems. In this situation time reversal invariance implies Kramers' degeneracy [30], i.e. all eigenvalues have at least multiplicity two, which can be seen as follows [31]. If ψ is an eigenfunction of $\hat{H}$ so is $\hat{T}\psi$ (since $\hat{T}$ and $\hat{H}$ commute and the eigenvalues of $\hat{H}$ are real). However, one finds that ψ and $\hat{T}\psi$ are orthogonal since

$$\langle\psi, \hat{T}\psi\rangle = \overline{\langle\hat{T}\psi, \hat{T}^2\psi\rangle} = -\overline{\langle\hat{T}\psi, \psi\rangle} = -\langle\psi, \hat{T}\psi\rangle = 0 . \tag{7.4}$$

Thus, all eigenvalues of $\hat{H}$ are degenerate. When discussing spectral statistics for time reversal invariant systems with half-integer spin we have to remove this degeneracy "by hand" before we can compare to the respective results from random matrix theory. Finally, we remark that any anti-unitary operator can by written as the product of a unitary operator and $\hat{K}$, the operator of complex conjugation, see e.g. [32]. Invariance with respect to an anti-unitary operator other than $\hat{T}$ is called non-conventional time reversal invariance [15]. As explained in the preceeding remark it can be treated in the same way as conventional time reversal. It is only important to distinguish between the cases where the anti-unitary operator squares to plus or minus unity. A situation in which non-conventional time reversal plays a role and which is similar to the problems discussed in the following section was investigated in [33].

Given a suitably desymmetrised system we can ask on which scale its spectral statistics should be analysed. The mean density of states is given by Weyl's law (classically accessible phase space volume divided by $(2\pi\hbar)^d$, the volume of Planck–cell) and is system specific. However, when talking about spectral statistics we always refer to the fluctuations about this mean behaviour. Therefore, we have to rescale the eigenvalues such that their mean separation is unity. This process is known as unfolding. Consider an energy spectrum $\{E_n\}$ with mean density $\bar{d}(E) = |\Omega_E|/(2\pi\hbar)^d$, which we have extracted from the non-oscillating part of the trace formula. Considering spectral fluctuations in the vicinity of a fixed energy E we write $\bar{d} := \bar{d}(E)$ and define the unfolded spectrum $\{x_n\}$ by

$$x_n = \bar{d}E_n \ . \tag{7.5}$$

The spectral density of the unfolded spectrum now reads

$$d(x') := \sum_n \delta(x' - x_n) = \sum_n \delta(\bar{d}(E' - E_n)) = d(E')/\bar{d} \ , \tag{7.6}$$

where we have defined $x' := \bar{d}E'$. Since the mean density of the original spectrum was $\bar{d}(E')$, for E' close to our reference energy E we have obtained a spectrum with mean density one. As we will see in Sect. 7.3 the unfolding (7.5) is particularly convenient in a semiclassical context, in which we can ensure that in the semiclassical limit $\hbar \to 0$ infinitely many eigenvalues condense in a small vicinity of the energy E.

We remark that in numerical studies one usually has to deal with a finite part of the spectrum, for which, if it is sufficiently close to the asymptotic regime, one expects to find generic spectral statistics. Then a local unfolding procedure like (7.5) is inconvenient. Instead one might choose to use either $x'_n := \bar{d}(E_n)E_n$ or $x''_n := \bar{N}(E_n)$ with $\bar{N}(E) = \int^E \bar{d}(E')\,\mathrm{d}E'$. Asymptotically all three prescriptions are equivalent and, in particular, lead to an unfolded spectrum with mean density one.

7.2 Time Reversal Invariance in the Trace Formula

We show that time reversal invariance of a quantum system gives rise to identical contributions of periodic orbits and their time reversed partners in the trace formula.

Time reversal invariance, $[\hat{H}, \hat{T}] = 0$, is equivalent to

$$\hat{T}\hat{H}\hat{T}^{-1} = \hat{H} \ . \tag{7.7}$$

For a Pauli Hamiltonian of the form, cf. 3.1,

$$\hat{H}_{\mathrm{P}} = H(\hat{\boldsymbol{p}}, \hat{\boldsymbol{x}}) + \hat{\boldsymbol{s}}\boldsymbol{\mathcal{B}}(\hat{\boldsymbol{p}}, \hat{\boldsymbol{x}}) \ , \tag{7.8}$$

where by $f(\hat{\boldsymbol{p}}, \hat{\boldsymbol{x}})$ we mean the Weyl quantisation of the phase space function $f(\boldsymbol{p}, \boldsymbol{x})$, we have

$$\hat{T}\hat{H}_{\mathrm{P}}\hat{T}^{-1} = H(-\hat{\boldsymbol{p}}, \hat{\boldsymbol{x}}) - \hat{\boldsymbol{s}}\boldsymbol{\mathcal{B}}(-\hat{\boldsymbol{p}}, \hat{\boldsymbol{x}}) \ , \tag{7.9}$$

i.e. time reversal invariance implies

$$H(-\boldsymbol{p}, \boldsymbol{x}) = H(\boldsymbol{p}, \boldsymbol{x}) \quad \text{and} \quad \boldsymbol{\mathcal{B}}(-\boldsymbol{p}, \boldsymbol{x}) = -\boldsymbol{\mathcal{B}}(\boldsymbol{p}, \boldsymbol{x}) \ . \tag{7.10}$$

Similarly, for a Dirac Hamiltonian

$$\hat{H}_{\mathrm{D}} = c\boldsymbol{\alpha}\left(\hat{\boldsymbol{p}} - \frac{e}{c}\boldsymbol{A}\right) + \beta mc^2 + e\phi \tag{7.11}$$

we find

$$\hat{T}\hat{H}_{\mathrm{D}}\hat{T}^{-1} = c(-\boldsymbol{\alpha})\left(-\hat{\boldsymbol{p}} - \frac{e}{c}\boldsymbol{A}\right) + \beta mc^2 + e\phi \ , \tag{7.12}$$

i.e. $[\hat{H}, \hat{T}] = 0 \Leftrightarrow \boldsymbol{A} \equiv 0$. As in eq. (7.10) this implies

$$H^{\pm}(-\boldsymbol{p}, \boldsymbol{x}) = H^{\pm}(\boldsymbol{p}, \boldsymbol{x}) \quad \text{and} \quad \boldsymbol{\mathcal{B}}^{\pm}(-\boldsymbol{p}, \boldsymbol{x}) = -\boldsymbol{\mathcal{B}}^{\pm}(\boldsymbol{p}, \boldsymbol{x}) \ . \tag{7.13}$$

cf. eqs. (3.186) and (3.199).

A classical system is invariant under time reversal if the equations of motion do not change under the transformation $(\boldsymbol{p}, \boldsymbol{x}, t) \mapsto (-\boldsymbol{p}, \boldsymbol{x}, -t)$. Thus we must consider the mapped Hamilton's equations of motion,

$$\begin{Bmatrix} \dot{\boldsymbol{x}} = \nabla_{\boldsymbol{p}} H(\boldsymbol{p}, \boldsymbol{x}) \\ \dot{\boldsymbol{p}} = -\nabla_{\boldsymbol{x}} H(\boldsymbol{p}, \boldsymbol{x}) \end{Bmatrix} \quad \mapsto \quad \begin{Bmatrix} -\dot{\boldsymbol{x}} = -\nabla_{\boldsymbol{p}} H(-\boldsymbol{p}, \boldsymbol{x}) \\ \dot{\boldsymbol{p}} = -\nabla_{\boldsymbol{x}} H(-\boldsymbol{p}, \boldsymbol{x}) \end{Bmatrix} \ . \tag{7.14}$$

Clearly these are invariant if and only if $H(-\boldsymbol{p}, \boldsymbol{x}) = H(\boldsymbol{p}, \boldsymbol{x})$. Thus quantum mechanical time reversal invariance implies time reversal invariance of the corresponding classical translational dynamics. In terms of the corresponding flow ϕ_H^t this property can be expressed in the following way. Denote by τ the map

$$\tau : (\boldsymbol{p}, \boldsymbol{x}) \mapsto (-\boldsymbol{p}, \boldsymbol{x}) \ . \tag{7.15}$$

Then time reversal invariance means that

$$\tau \circ \phi_H^t \circ \tau \circ \phi_H^t = id \, , \tag{7.16}$$

i.e. if we propagate up to time t, reverse the momentum, propagate for another time interval of length t with the same equations of motion, and finally reverse the momentum again, we end up at the initial point. Notice that eq. (7.16) is equivalent to

$$\tau \circ \phi_H^t = \phi_H^{-t} \circ \tau \, . \tag{7.17}$$

For the trace formula this immediately implies that, in case of time reversal invariant systems, for each manifold $\mathcal{M}_s$ of periodic points, there is a time reversed parter $\tau\mathcal{M}_s$ with $T_{\tau\mathcal{M}_s} = T_{\mathcal{M}_s}$ and $S_{\tau\mathcal{M}_s} = S_{\mathcal{M}_s}$. In addition to the period $T_{\mathcal{M}_s}$ and the action $S_{\mathcal{M}_s}$ all spin-independent terms in the amplitudes $\mathcal{A}_{\mathcal{M}_s}$ are also invariant.

For the spin part we need to consider the equation of spin precession which, by the same transformation as above, is mapped as

$$\dot{\boldsymbol{s}} = \boldsymbol{\mathcal{B}}\left(\phi_H^t(\boldsymbol{p},\boldsymbol{x})\right) \times \boldsymbol{s} \quad \mapsto \quad -\dot{\boldsymbol{s}} = \boldsymbol{\mathcal{B}}\left(\phi_H^{-t}(-\boldsymbol{p},\boldsymbol{x})\right) \times \boldsymbol{s} \, . \tag{7.18}$$

For time reversal invariant systems we can use (7.17) and (7.10) in order to show that

$$\begin{aligned} \boldsymbol{\mathcal{B}}\left(\phi_H^{-t}(-\boldsymbol{p},\boldsymbol{x})\right) &= \boldsymbol{\mathcal{B}}\left((\phi_H^{-t} \circ \tau)(\boldsymbol{p},\boldsymbol{x})\right) = \boldsymbol{\mathcal{B}}\left((\tau \circ \phi_H^t)(\boldsymbol{p},\boldsymbol{x})\right) \\ &= -\boldsymbol{\mathcal{B}}\left(\phi_H^t(\boldsymbol{p},\boldsymbol{x})\right) \, , \end{aligned} \tag{7.19}$$

i.e. in the mapped equation of spin precession (7.18) the minus signs cancel. One might wonder why there was no need to include a map $\boldsymbol{s} \mapsto -\boldsymbol{s}$ into our considerations; the answer is that the special form of the equation of spin precession renders such a transformation irrelevant.

Finally we conclude that quantum mechanical time reversal invariance, $[\hat{H}, \hat{T}] = 0$, implies the classical time reversal invariance of the skew product flow Y_{cl}^t. Thus, in this situation the contributions to the trace formula of a manifold $\mathcal{M}_s$ of periodic orbits and its time reversed partner $\tau\mathcal{M}_s$ are identical.

7.3 Spectral Two-Point Form Factor

We discuss the definition of the spectral form factor of a quantum system and in which sense we can expect convergence towards universal form factors from random matrix theory or the two-point form factor of a Poisson process.

Consider the eigenvalues E_n of a quantum Hamiltonian in an interval

$$I(E,\hbar) := [E - \hbar^a\omega, E + \hbar^a\omega] \tag{7.20}$$

centered about an energy E. The parameter $\omega = const.$ shall be positive. and the exponent a will be chosen such that in the semiclassical limit $\hbar \to 0$ the length of the interval shrinks to zero, i.e.

$$a > 0\,. \tag{7.21}$$

Thus, we may unfold the eigenvalues E_n using the mean density at energy E rather than E_n , i.e. we define

$$x_n := \bar{d}\, E_n\,, \tag{7.22}$$

with $\bar{d} := \bar{d}(E)$ fixed, cf. (7.5). The spectral density $\rlap{/}d(x')$, $x' = \bar{d}E'$, of the unfolded spectrum is then given by

$$\rlap{/}d(x') := \sum_n \delta(x' - x_n) = \frac{1}{\bar{d}} \sum_n \delta(E' - E_n) = d(E')/\bar{d}\,. \tag{7.23}$$

From Weyl's law, which appears as the period-zero term in the trace formula, we have

$$\bar{d} = (2s+1)\frac{|\Omega_E|}{(2\pi\hbar)^d} \tag{7.24}$$

for a system with d translational degrees of freedom and with spin s. In the semiclassical limit the number N_I of eigenvalues in the interval $I(E,\hbar)$ can be estimated by its length and the mean density,

$$N_I \sim 2\hbar^a \omega \bar{d}\,. \tag{7.25}$$

Since we do not want to analyse a finite spectrum we have to make sure that $N_I \to \infty$ as $\hbar \to 0$ although the length of the interval shrinks. This can be achieved by choosing

$$0 < a < d\,. \tag{7.26}$$

We now define the spectral two-point correlation function for the energies in the interval $I(E,\hbar)$,

$$R_2^I(s) := \left\langle \rlap{/}d\left(x' + \frac{s}{2}\right) \rlap{/}d\left(x' - \frac{s}{2}\right) \right\rangle_{x'} - 1\,, \tag{7.27}$$

where the spectral average is given by

$$\langle \cdot \rangle_{x'} := \frac{1}{2\hbar^a \omega \bar{d}} \int_{x - \hbar^a \omega \bar{d}}^{x + \hbar^a \omega \bar{d}} \cdot \, \mathrm{d}x'\,, \tag{7.28}$$

with $x = \bar{d}E$. Notice that there may be a finite number of eigenvalues outside $I(E,\hbar)$ contributing to $R_2^I(s)$. However, due to the accumulation of infinitely many eigenvalues in the semiclassical limit, asymptotically this will not influence the statistics and by substituting (7.23) and (7.25) we can perform the x'-integration, yielding

$$R_2^I(s) \sim \frac{1}{N_I} \sum_{n,n'=1}^{N_I} \delta(x_n - x_{n'} - s) - 1 \,, \quad \hbar \to 0 \,. \tag{7.29}$$

By means of (7.23) we can also rewrite $R_2^I(s)$ in terms of the original spectral density $d(E)$ as follows,

$$R_2^I(s) = \left\langle \frac{d(E' + s/(2\bar{d}))\, d(E' - s/(2\bar{d}))}{\bar{d}^2} \right\rangle_{E'} - 1 \,, \tag{7.30}$$

where by a change of variables the average is now given by

$$\langle \cdot \rangle_{E'} := \frac{1}{2\hbar^a \omega} \int_{E-\hbar^a\omega}^{E+\hbar^a\omega} \cdot \, \mathrm{d}E' \,. \tag{7.31}$$

The spectral form factor is the Fourier transform of $R_2^I(s)$, i.e.

$$K_I(\tau) := \int_{\mathbb{R}} R_2^I(s)\, \mathrm{e}^{-2\pi i \tau s}\, \mathrm{d}s \,. \tag{7.32}$$

Both the form factor and the two-point correlation function are distributions which need to be evaluated on suitable test functions. For instance, with an arbitrary function $\phi \in \mathcal{S}(\mathbb{R})$ we have to investigate

$$\langle K_I, \phi \rangle := \int_{\mathbb{R}} K_I(\tau)\, \phi(\tau)\, \mathrm{d}\tau = \int_{\mathbb{R}} R_2^I(s)\, \hat{\phi}(2\pi s)\, \mathrm{d}s \,, \tag{7.33}$$

where we have exploited the symmetry of the two-point correlation function, $R_2^I(-s) = R_2^I(s)$. Recall that our convention for the Fourier transform is

$$\hat{\phi}(k) = \int_{\mathbb{R}} \phi(x)\, \mathrm{e}^{ikx}\, \mathrm{d}x \quad \text{and} \quad \phi(x) = \frac{1}{2\pi} \int_{\mathbb{R}} \hat{\phi}(k)\, \mathrm{e}^{-ikx}\, \mathrm{d}k \,, \tag{7.34}$$

cf. (3.16) in Sect. 3.2. When we say that $K_I(\tau)$ converges to some limiting function $K(\tau)$ as $\hbar \to 0$, we mean that

$$\lim_{\hbar \to 0} \int_{\mathbb{R}} K_I(\tau)\, \phi(\tau)\, \mathrm{d}\tau = \int_{\mathbb{R}} K(\tau)\, \phi(\tau)\, \mathrm{d}\tau \tag{7.35}$$

for any ϕ in a suitable class of test functions. If the quantum system under consideration is the quantisation of a chaotic classical system, according to the BGS-conjecture, we expect that $K_I(\tau)$ converges to a form factor from random matrix theory. For systems which are not invariant under time reversal the relevant random matrix ensemble is the GUE, for which the form factor reads [7][2]

$$K_{\mathrm{GUE}}(\tau) = \begin{cases} \tau \,, & 0 \le \tau \le 1 \\ 1 \,, & \tau > 1 \end{cases} \,. \tag{7.36}$$

[2] Our notation $K(\tau)$ is related to the notation $b(k)$ in [7] by $K(\tau) = 1 - b(\tau)$.

We remark that due to the symmetry of the two-point correlation function the form factor is also symmetric, $K(-\tau) = K(\tau)$, and thus it is sufficient to to state the results for positive τ. For time reversal invariant systems with integer spin we expect GOE-statistics, in which case the form factor is given by

$$K_{\mathrm{GOE}}(\tau) = \begin{cases} 2\tau - \tau \ln(1+2\tau) \ , & 0 \leq \tau \leq 1 \\ 2 - \tau \ln\left(\frac{2\tau+1}{2\tau-1}\right) \ , & \tau > 1 \end{cases} . \tag{7.37}$$

Finally, if we are dealing with half-integer spin and invariance under time reversal, we have to compare with GSE statistics with the form factor

$$K_{\mathrm{GSE}}(\tau) = \begin{cases} \frac{1}{2}\tau - \frac{1}{4}\tau \ln|1-\tau| \ , & 0 \leq \tau \leq 2 \\ 1 \ , & \tau > 2 \end{cases} . \tag{7.38}$$

What will be accessible for semiclassical considerations is the regime of small τ in which the random matrix form factors behave like

$$K(\tau) = \begin{cases} \tau \ , & \mathrm{GUE} \\ 2\tau \ , & \mathrm{GOE} \\ \frac{1}{2}\tau \ , & \mathrm{GSE} \end{cases} + \mathcal{O}\left(\tau^2\right) . \tag{7.39}$$

The behaviour of $K(\tau)$ for large arguments is solely determined by the spectrum being discrete [17, 34]. In order to see this one observes that in the absence of systematic degeneracies the strongest singularity of $R_2(s)$ is located at $s = 0$, where in (7.29) only the terms with $n = n'$ contribute, and thus the two-point correlation function behaves like

$$R_2(s) \sim \delta(s) \ , \quad s \to 0 . \tag{7.40}$$

This singularity in turn determines the behaviour of the Fourier transform for large arguments, i.e. we conclude that

$$K(\tau) \sim \int \delta(s)\, \mathrm{e}^{-2\pi i \tau s}\, \mathrm{d}s = 1 \ , \quad \tau \to \infty . \tag{7.41}$$

The form factor of a Poisson process is determined solely by this behaviour in the whole τ-range, since there are no correlations in the spectrum, i.e.

$$K_{\mathrm{Poisson}} \equiv 1 . \tag{7.42}$$

This behaviour is expected for integrable systems [6]. Recently additional "universality" classes have been conjectured to play a role for so-called pseudo-integrable systems. There one observes intermediate spectral statistics. In [35] a class of ensembles was introduced for which the form factors, unlike the standard random matrix form factors, do not vanish at $\tau = 0$ but have a value smaller than one. The most prominent example is the so-called semi-Poisson ensemble for which the form factor reads

$$K_{\mathrm{sP}}(\tau) = 1 - \frac{2}{4+\pi^2\tau^2} = \frac{1}{2} + \mathcal{O}\left(\tau^2\right) , \quad \tau \to 0 . \tag{7.43}$$

A possible connection with these ensembles will be discussed in Sects. 7.3.4 and 7.4, in the context of systems for which the translational dynamics is integrable but the skew product is not.

7.3.1 Diagonal Approximation

We derive a semiclassical representation of the spectral form factor in terms of regularised trace formulae. Then we motivate the so-called diagonal approximation [16, 17], which is the basis of a semiclassical analysis of spectral statistics.

For the definition of the form factor or the two-point correlation function we need the spectral density of our system. In various places we have remarked that a semiclassical representation of the spectral density can be obtained from the regularised trace formula by formally replacing the test function $\rho(\omega)$ with $\delta(\hbar\omega)$. This is indeed how Gutzwiller originally stated his trace formula [36]. However, in order to circumvent various convergence problems, we prefer to work with the regularised trace formula. To this end we introduce a limiting procedure which will allow us to work with convergent expressions and only at the end remove the regularisation; see [29] for a similar treatment.

First recall that for a smooth function $\eta \in \mathcal{S}(\mathbb{R})$ whose Fourier transform has compact support, $\hat{\eta} \in C_0^\infty(\mathbb{R})$, which we choose to be normalised as

$$\int_{\mathbb{R}} \eta(\omega)\, \mathrm{d}\omega = 1 , \tag{7.44}$$

we can define a series of functions which converge to the δ-function by

$$\rho_\varepsilon(\omega) := \frac{1}{\varepsilon}\, \eta\Big(\frac{\omega}{\varepsilon}\Big) . \tag{7.45}$$

Obviously, ρ_ε is normalised for all $\varepsilon > 0$ and for $\varepsilon \to 0$ converges to $\delta(\omega)$. Thus, if in the trace formula we replace $\rho(\omega)$ by $\rho_\varepsilon(\omega)/\hbar = \eta(\omega/\varepsilon)/\hbar\varepsilon$, we obtain a smoothed density of states,

$$d_\varepsilon(E) := \sum_n \frac{1}{\hbar\varepsilon}\, \eta\Big(\frac{E_n - E}{\hbar\varepsilon}\Big) , \tag{7.46}$$

for which we have a convergent trace formula for all $\varepsilon > 0$ and which in the limit $\varepsilon \to 0$ becomes the spectral density $d(E)$. On the right hand side of the trace formula (3.67) we have to replace $\hat{\rho}(t)$ by

$$\int_{\mathbb{R}} \frac{1}{\hbar\varepsilon}\, \eta\Big(\frac{\omega}{\varepsilon}\Big)\, \mathrm{e}^{\mathrm{i}\omega t}\, \mathrm{d}\omega = \frac{1}{\hbar}\, \hat{\eta}(\varepsilon t) . \tag{7.47}$$

Thus, we obtain the following trace formula for a smoothed spectral density,

$$\begin{aligned} d_\varepsilon(E) &= (2s+1)\frac{|\Omega_E|}{(2\pi\hbar)^d}\,\hat{\eta}(0)\,(1+\mathcal{O}(\hbar)) \\ &\quad + \sum_{\mathcal{M}_s} \frac{\hat{\eta}(\varepsilon T_{\mathcal{M}_s})}{2\pi\hbar}\,\mathcal{A}_{\mathcal{M}_s}\,\mathrm{e}^{(\mathrm{i}/\hbar)S_{\mathcal{M}_s}}\,(1+\mathcal{O}(\hbar))\,. \end{aligned} \tag{7.48}$$

Notice that, due to the normalisation (7.44), $\hat{\eta}(0) = 1$ and thus the Weyl term gives rise to the mean spectral density

$$\overline{d}(E) = (2s+1)\frac{|\Omega_E|}{(2\pi\hbar)^d}\,(1+\mathcal{O}(\hbar))\,, \tag{7.49}$$

independent of the smoothing. The fluctuating part of $d_\varepsilon(E)$ will be denoted by

$$d_\varepsilon^{\mathrm{fl}}(E) := \sum_{\mathcal{M}_s} \frac{\hat{\eta}(\varepsilon T_{\mathcal{M}_s})}{2\pi\hbar}\,\mathcal{A}_{\mathcal{M}_s}\,\mathrm{e}^{(\mathrm{i}/\hbar)S_{\mathcal{M}_s}}\,(1+\mathcal{O}(\hbar))\,. \tag{7.50}$$

Here we explicitly see the clash of limits if we want to write down a trace formula for $d(E)$ itself. Although for all $\varepsilon > 0$ the finite support of $\hat{\eta}$ guarantees a convergent periodic orbit sum, in the limit $\varepsilon \to 0$ the cut-off is removed and contributions of arbitrarily long orbits have to be included. This was already apparent from the formal replacement $\rho(\omega) \to \delta(\hbar\omega)$ which completely removes the cut-off from the periodic orbit sum, but now we have a smooth way to approach this limit. The following discussion will always be carried out for non-vanishing ε, only in the end taking the limit $\varepsilon \to 0$.

Some expressions will not be well defined without the cut-off. When encountering this problem, we will have to leave the cut-off in place and try to still extract the required information on the distribution of levels. This can be achieved as follows. We are interested in correlations on the scale of the mean level spacing $1/\overline{d}$. Thus, if we denote by σ^2 the variance of η, we have the requirement

$$\varepsilon\hbar\sigma \approx \frac{1}{\overline{d}}\,. \tag{7.51}$$

By the uncertainty relation of the Fourier transform this implies a variance of $(2\pi/\sigma)^2$ for $\hat{\eta}$, thus enforcing an effective cut-off of the periodic orbit sum by

$$\varepsilon T_\gamma \lesssim \frac{2\pi}{\sigma} \quad \Leftrightarrow \quad T_\gamma \lesssim 2\pi\hbar\overline{d} =: T_H\,. \tag{7.52}$$

The time T_H which corresponds to the scale of the mean level spacing is known as the Heisenberg time.

We can now obtain a semiclassical representation of the two-point correlation function by inserting (7.48) into (7.27), i.e. we define

$$\begin{aligned}
R_2^{I,\varepsilon}(s) :=& \left\langle \frac{d_\varepsilon(E'+s/(2\bar{d}))\, d_\varepsilon(E'-s/(2\bar{d}))}{\bar{d}^2} \right\rangle_{E'} - 1 \\
=& \left\langle \frac{\bar{d}(E'+s/(2\bar{d}))\, \bar{d}(E'-s/(2\bar{d}))}{\bar{d}^2} \right\rangle_{E'} - 1 \\
&+ \left\langle \frac{\bar{d}(E'+s/(2\bar{d}))\, d_\varepsilon^{\text{fl}}(E'-s/(2\bar{d}))}{\bar{d}^2} \right\rangle_{E'} \\
&+ \left\langle \frac{d_\varepsilon^{\text{fl}}(E'+s/(2\bar{d}))\, \bar{d}_\varepsilon(E'-s/(2\bar{d}))}{\bar{d}^2} \right\rangle_{E'} \\
&+ \left\langle \frac{d_\varepsilon^{\text{fl}}(E'+s/(2\bar{d}))\, d_\varepsilon^{\text{fl}}(E'-s/(2\bar{d}))}{\bar{d}^2} \right\rangle_{E'} .
\end{aligned} \tag{7.53}$$

The first term can be simplified by expanding the mean density about the center of $I(E,\hbar)$,

$$\bar{d}\left(E' \pm \frac{s}{2\bar{d}}\right) = \bar{d}(E) + \mathcal{O}\left(E' \pm \frac{s}{2\bar{d}} - E\right) = \bar{d} + \mathcal{O}\left(\hbar^a\right) , \tag{7.54}$$

where the last equation holds because $\mathcal{O}\left(E'-E\right) = \mathcal{O}\left(\hbar^a\right)$ due to the choice of $I(E,\hbar)$ and because $\bar{d} = \mathcal{O}\left(\hbar^{-d}\right)$. Thus we have

$$\left\langle \frac{\bar{d}(E'+s/(2\bar{d}))\, \bar{d}(E'-s/(2\bar{d}))}{\bar{d}^2} \right\rangle_{E'} = 1 + \mathcal{O}\left(\hbar^{d+a}\right) . \tag{7.55}$$

The second and the third term in (7.53) vanish asymptotically because we average the function $d_\varepsilon^{\text{fl}}$ which fluctuates about zero. Thus we are left with

$$R_2^{I,\varepsilon}(s) \sim \left\langle \frac{d_\varepsilon^{\text{fl}}(E'+s/(2\bar{d}))\, d_\varepsilon^{\text{fl}}(E'-s/(2\bar{d}))}{\bar{d}^2} \right\rangle_{E'} , \tag{7.56}$$

which is the energy average of a double periodic orbit sum. Explicitly inserting the trace formula (7.48) we have

$$\begin{aligned}
R_2^{I,\varepsilon}(s) \sim \Bigg\langle & \frac{1}{T_H^2} \sum_{\mathcal{M}_s} \sum_{\mathcal{M}_s'} \hat{\eta}(\varepsilon T_{\mathcal{M}_s})\, \hat{\eta}(\varepsilon T_{\mathcal{M}_s'})\, \mathcal{A}_{\mathcal{M}_s} \mathcal{A}_{\mathcal{M}_s'} \\
& \times \exp\left(\frac{\mathrm{i}}{\hbar}\left[S_{\mathcal{M}_s}(E'+s/(2\bar{d})) + S_{\mathcal{M}_s'}(E'-s/(2\bar{d}))\right]\right) \Bigg\rangle_{E'} ,
\end{aligned} \tag{7.57}$$

where we have suppressed the arguments of the slowly varying amplitudes $\mathcal{A}$ and introduced the Heisenberg time T_H, see (7.52). Recall that we can always split periodic orbit sums into sums over primitive orbits $\mathcal{M}_s{}^{\#}$ and their k-fold repetitions, $k \in \mathbb{Z} \setminus \{0\}$. Doing so in the $\mathcal{M}_s'$-sum, then changing variables from k' to $-k'$ and afterwards merging the sums again, we obtain

$$R_2^{I,\varepsilon}(s) \sim \left\langle \frac{1}{T_H^2} \sum_{\mathcal{M}_s} \sum_{\mathcal{M}'_s} \hat{\eta}(\varepsilon T_{\mathcal{M}_s}) \overline{\hat{\eta}(-\varepsilon T_{\mathcal{M}'_s})} \times \right.$$
$$\left. \times \mathcal{A}_{\mathcal{M}_s} \overline{\mathcal{A}_{\mathcal{M}'_s}} \, \mathrm{e}^{\frac{\mathrm{i}}{\hbar}(S_{\mathcal{M}_s}(E'+\frac{s}{2\bar{d}})-S_{\mathcal{M}'_s}(E'-\frac{s}{2\bar{d}}))} \right\rangle_{E'} , \tag{7.58}$$

where we have used that $k' \mapsto -k'$ implies $\mathcal{A}_{\mathcal{M}'_s} \mapsto \overline{\mathcal{A}_{\mathcal{M}'_s}}$, i.e. the spectral density $d(E)$ is real. We may now expand the exponent in the same way as we expanded the mean density in eq. (7.54),

$$\begin{aligned} &S_{\mathcal{M}_s}\left(E' + \frac{s}{2\bar{d}}\right) - S_{\mathcal{M}'_s}\left(E' - \frac{s}{2\bar{d}}\right) \\ &= S_{\mathcal{M}_s} - S_{\mathcal{M}'_s} + (T_{\mathcal{M}_s} - T_{\mathcal{M}'_s})(E' - E) + \frac{T_{\mathcal{M}_s} + T_{\mathcal{M}'_s}}{2} \frac{s}{\bar{d}} + \mathcal{O}(\hbar^a) , \end{aligned} \tag{7.59}$$

where we have omitted energy arguments whenever the energy equals E. We now observe a rapidly oscillating term inside the energy average of (7.59) which reads

$$\begin{aligned} &\left\langle \exp\left(\frac{\mathrm{i}}{\hbar}(T_{\mathcal{M}_s} - T_{\mathcal{M}'_s})(E' - E)\right)\right\rangle_{E'} \\ &= \frac{1}{2\hbar^a\omega} \int_{-\hbar^a\omega}^{+\hbar^a\omega} \mathrm{e}^{(\mathrm{i}/\hbar)(T_{\mathcal{M}_s} - T_{\mathcal{M}'_s})E''} \, \mathrm{d}E'' \\ &= \begin{cases} 1 & , T_{\mathcal{M}_s} = T_{\mathcal{M}'_s} \\ \dfrac{\sin((T_{\mathcal{M}_s} - T_{\mathcal{M}'_s})\hbar^{a-1}\omega)}{(T_{\mathcal{M}_s} - T_{\mathcal{M}'_s})\hbar^{a-1}\omega} & , T_{\mathcal{M}_s} \neq T_{\mathcal{M}'_s} \end{cases} . \end{aligned} \tag{7.60}$$

We see that terms with a large difference

$$\delta T := T_{\mathcal{M}_s} - T_{\mathcal{M}'_s} \tag{7.61}$$

are suppressed, and one might ask whether there is a regime in which the double sum is solely determined by the diagonal terms, i.e. those terms with $\delta T = 0$. However, in general we are not able to give a lower bound for $\delta T \neq 0$ and thus are unable to estimate the contribution of the non-diagonal terms. Only a rough estimate can be obtained as follows. Denote by $\mathcal{N}(T)$ the number of periodic orbits with period less than T, i.e.

$$\mathcal{N}(T) = \#\{T_{\mathcal{M}_s} \,|\, T_{\mathcal{M}_s} < T\} . \tag{7.62}$$

Then the density of periods in the vicinity of T is given by $\mathrm{d}\mathcal{N}(T)/\mathrm{d}T$ and we can expect a typical δT to be approximately given by

$$\delta T \approx \left[\frac{\mathrm{d}\mathcal{N}(T)}{\mathrm{d}T}\right]^{-1} . \tag{7.63}$$

For instance in the case of hyperbolic systems we have $\mathcal{N}(T) = \mathcal{O}(\exp(hT))$, where $h > 0$ is the topological entropy. For integrable systems we typically have an algebraically growing $\mathcal{N}(T)$.

Since (7.63) is not the lower bound one would need in order to make definite statements but just a rough estimate of the order of magnitude, we refrain from a more detailed discussion. Instead we conclude that we may at the best expect the double sum in (7.57) to be dominated be the diagonal terms, if we can control the magnitude of the periods of the longest orbits which contribute. In eq. (7.57) the only restriction on the periods is given by our condition that the test function $\hat{\eta}$ be compactly supported. However, in the end we want to take the limit $\varepsilon \to 0$, thus removing the cut-off from the periodic orbit sums and possibly generating arbitrarily small differences δT. Therefore, we can not use the cut-off provided by $\hat{\eta}$ in order to justify the neglect of non-diagonal terms.

At this point it is helpful to recall that the two-point correlation function is a distribution which should be evaluated on a test function. By doing so, and expressing the result in terms of the spectral form factor $K(\tau)$ we will get a restriction on the τ-range in which we can invoke the diagonal approximation. Inserting (7.57) into (7.33) we obtain

$$\langle K_{I,\varepsilon}, \phi \rangle \sim \left\langle \frac{1}{T_H^2} \sum_{\mathcal{M}_s} \sum_{\mathcal{M}_s'} \mathcal{A}_{\mathcal{M}_s} \overline{\mathcal{A}_{\mathcal{M}_s'}}\, \phi\left(\frac{T_{\mathcal{M}_s} + T_{\mathcal{M}_s'}}{2T_H} \right) \right. \\ \left. \times \exp\left(\frac{\mathrm{i}}{\hbar} [S_{\mathcal{M}_s} - S_{\mathcal{M}_s'} + (T_{\mathcal{M}_s} - T_{\mathcal{M}_s'})(E' - E)] \right) \right\rangle_{E'} . \tag{7.64}$$

Notice that since $\overline{d} = \mathcal{O}(\hbar^{-d})$ in the semiclassical limit the Heisenberg time goes to infinity, $T_H \to \infty$. We want to investigate terms with $T_{\mathcal{M}_s} \neq T_{\mathcal{M}_s'}$, $T_{\mathcal{M}_s} \approx T_{\mathcal{M}_s'}$, since, according to eq. (7.60), these yield the largest non-diagonal contributions. For a compactly supported test function ϕ, say

$$\operatorname{supp} \phi \subseteq [-\tau^*, \tau^*] , \tag{7.65}$$

the longest orbits contributing have lengths

$$T_{\mathcal{M}_s} \approx \tau^* T_H . \tag{7.66}$$

In order for the diagonal approximation to be good, τ^* has to vanish fast enough in the semiclassical limit. Hence the diagonal form factor provides us, at the best, with an estimate for the spectral form factor for small arguments. We can read off the diagonal form factor from (7.64) to

$$K_{\text{diag}}(\tau) := \frac{1}{T_H^2} \sum_{\mathcal{M}_s} \sum_{\substack{\mathcal{M}_s' \\ T_{\mathcal{M}_s'} = T_{\mathcal{M}_s}}} \mathcal{A}_{\mathcal{M}_s} \overline{\mathcal{A}_{\mathcal{M}_s'}}\, \mathrm{e}^{(\mathrm{i}/\hbar)[S_{\mathcal{M}_s} - S_{\mathcal{M}_s'}]}\, \delta\left(\tau - \frac{T_{\mathcal{M}_s}}{T_H} \right) , \tag{7.67}$$

where we have already performed the E'-average and taken the limit $\varepsilon \to 0$. In all cases of interest, which will be discussed in the following sections, $T_{\mathcal{M}_s} = T_{\mathcal{M}'_s}$ will imply $S_{\mathcal{M}_s} = S_{\mathcal{M}'_s}$ and $\mathcal{A}_{\mathcal{M}_s} = \mathcal{A}_{\mathcal{M}'_s}$, cf. the discussion in Sect. 7.2. Therefore, if we denote by $g_{\mathcal{M}_s}$ the number of periodic orbits $\mathcal{M}'_s$ with $T_{\mathcal{M}'_s} = T_{\mathcal{M}_s}$, the diagonal form factor finally reads

$$K_{\text{diag}}(\tau) = \frac{1}{T_H} \sum_{\mathcal{M}_s} g_{\mathcal{M}_s} |\mathcal{A}_{\mathcal{M}_s}|^2 \, \delta(\tau T_H - T_{\mathcal{M}_s}) \,. \tag{7.68}$$

We will now use information on the classical dynamics, namely the sum rules for the skew product Y^t_{cl}, in order to show that the diagonal form factor coincides with the small-τ asymptotics of the universal form factors listed in Sect. 7.3.

7.3.2 Chaotic Systems

First we want to discuss strongly chaotic systems for which all periodic orbits are isolated and unstable. This requirement concerns the translational dynamics only; a condition on the spin dynamics will be imposed later in this section.

In the case of only isolated and unstable periodic orbits we have, cf. (3.103),

$$\mathcal{A}_{\mathcal{M}_s} \equiv \mathcal{A}_\gamma = \frac{T^\#_\gamma \, \chi_s(\alpha_\gamma) \, \mathrm{e}^{-\mathrm{i}(\pi/2)\mu_\gamma}}{|\det(\mathbb{M}_\gamma - \mathbb{1}_{2d-2})|^{1/2}} \,. \tag{7.69}$$

We expect that for a generic system there are no systematic degeneracies of periods of periodic orbits. Thus, almost all orbits share the same value of $g_\gamma = g$ which is determined by symmetry alone. Since all unitary symmetries have to be removed before analysing spectral statistics, see Sect. 7.1, the only symmetry we have to consider is (anti-unitary) time reversal. Therefore, if the system is not invariant under time reversal, we have $g = 1$ and for time reversal invariant systems we have to set $g = 2$. For the latter to hold, we have to assume that the fraction of self-retracing orbits, i.e. orbits that are mapped onto themselves under time reversal, can be neglected. Then the diagonal form factor for strongly chaotic systems reads

$$K_{\text{diag}}(\tau) = \frac{g}{T_H} \sum_\gamma \frac{T^{\#2}_\gamma \chi^2_s(\alpha_\gamma)}{|\det(\mathbb{M}_\gamma - \mathbb{1}_{2d-2})|} \, \delta(\tau T_H - T_\gamma) \,. \tag{7.70}$$

For a similar periodic orbit sum we have derived a sum rule in Sect. 6.5 under the condition that Y^t_{cl} be ergodic. For the translational dynamics this is a much weaker condition than the strong chaotic properties demanded before. However, for the combined dynamics of translational motion and spin precession we had not previously demanded any degree of chaoticity. Ergodicity of Y^t_{cl} is our only requirement here, in which case we have the relation (6.64),

$$\sum_\gamma \frac{T_\gamma^\# \chi_s^2(\alpha_\gamma)}{|\det(\mathbb{M}_\gamma - \mathbb{1}_{2d-2})|}\,\delta(T - T_\gamma) \sim 1\,, \quad T \to \infty\,. \tag{7.71}$$

Since in strongly chaotic systems the number of primitive periodic orbits grows exponentially, and since the contribution to (7.71) of the k-fold repetition of a primitive periodic orbit is damped exponentially by

$$\frac{1}{|\det(\mathbb{M}_\gamma - \mathbb{1}_{2d-2})|} \sim \mathrm{e}^{-u_\gamma k T_\gamma^\#}\,, \tag{7.72}$$

where u_γ is the largest stability exponent, the periodic orbit sum (7.71) is asymptotically dominated by the primitive periodic orbits. Upon multiplying with T we find

$$\sum_\gamma \frac{T_\gamma T_\gamma^\# \chi_s^2(\alpha_\gamma)}{|\det(\mathbb{M}_\gamma - \mathbb{1}_{2d-2})|}\,\delta(T - T_\gamma) \sim \sum_\gamma \frac{T_\gamma^{\#2} \chi_s^2(\alpha_\gamma)}{|\det(\mathbb{M}_\gamma - \mathbb{1}_{2d-2})|}\,\delta(T - T_\gamma) \sim T\,, \tag{7.73}$$

as $T \to \infty$. Thus, in the combined limit

$$T_H \to \infty\,, \quad \tau \to 0 \quad \text{and} \quad \tau T_H \to \infty\,, \tag{7.74}$$

the asymptotics of the diagonal form factor is given by

$$K_{\mathrm{diag}}(\tau) \sim g\tau\,. \tag{7.75}$$

For non-time reversal invariant systems ($g = 1$) this coincides with the small-τ asymptotics of $K_{\mathrm{GUE}}(\tau)$, see (7.39), whereas for time reversal invariant systems ($g = 2$) it is consistent with $K_{\mathrm{GOE}}(\tau)$. However, eq. (7.75) was derived for arbitrary spin $s = 0, 1/2, 1, \ldots$, and for time reversal invariant systems with half integer spin we do not expect statistics according to the GOE but according to the GSE. In order to understand this ostensive contradiction we have to remember that in the latter situation time reversal symmetry causes Kramers' degeneracy, i.e. all eigenvalues have multiplicity two. Spectral statistics should be discussed for the spectrum with this degeneracy removed, whereas the trace formula describes the full quantum spectrum and thus (7.75) holds for the degenerate spectrum. If we define a new spectrum by

$$\tilde{E}_n := E_{2n}\,, \tag{7.76}$$

its spectral density $\tilde{d}(E)$ and mean spectral density $\bar{\tilde{d}}$ are simply given by

$$\tilde{d}(E) = d(E)/2 \quad \text{and} \quad \bar{\tilde{d}} = \bar{d}/2\,, \tag{7.77}$$

respectively. Thus, by (7.30), the two-point correlation function of the non-degenerate spectrum reads

$$\tilde{R}_2(s) = R_2(2s)\,, \tag{7.78}$$

and the form factor is given by

$$\tilde{K}(\tau) = \tfrac{1}{2}\, K\left(\tfrac{1}{2}\tau\right) \ . \tag{7.79}$$

Finally, using the asymptotics (7.75) with $g = 2$ we find

$$\tilde{K}_{\text{diag}}(\tau) \sim \tfrac{1}{2}\tau \ , \tag{7.80}$$

which is consistent with GSE statistics as we expected.

7.3.3 Integrable Systems

The opposite of the situation discussed in the preceeding section is that of completely integrable systems. By this characterisation we do not only mean integrability of the translational flow ϕ_H^t but also integrability of the skew product Y_{sc}^t, which implies the former.

In this situation the manifolds $\mathcal{M}_s$ of periodic points are given by rational tori characterised by winding numbers $\boldsymbol{m}$. From (4.73) we can read off the modulus squared of the amplitudes,

$$|\mathcal{A}_{\boldsymbol{m}}|^2 = \frac{(2\pi)^{2d}\,\chi_s^2(\alpha_{\boldsymbol{m}})}{(2\pi\hbar)^{d-1}\,|\det \mathbb{D}_{\boldsymbol{m}}|} \ . \tag{7.81}$$

Again the question arises whether there are different tori for which the periodic orbits have the same actions. As previously we shall assume that generically such a degeneracy can only be caused by symmetries of the system. Certainly, there are always symmetries in an integrable system generated by the conserved observables commuting with the Hamiltonian. However, if we consider the image of a periodic orbit under these symmetry transformations we simply obtain other periodic orbits lying on the same torus. Thus, these symmetries do not give rise to extra pre-factors in (7.68). There might still be discrete symmetries left in the system and we could have time reversal invariance. In all these cases the crucial question is whether a pair of orbits related by such a symmetry lies on the same torus or on different tori. In the former case they do account for the pre-factor $g_{\mathcal{M}_s}$ in (7.68). However, in case of orbits on different tori, the torus quantisation, which was discussed in detail in Chap. 5, will give rise to degenerate semiclassical eigenvalues. Such a degeneracy has to be removed in the process of desymmetrisation, cf. Sect. 7.1. Therefore, we conclude that for integrable systems eq. (7.68) applies with $g_{\mathcal{M}_s} = 1$ for (almost) all tori.

Now the diagonal form factor is given by

$$K_{\text{diag}}(\tau) = \frac{1}{T_H} \sum_{\boldsymbol{m}} \frac{(2\pi)^{d}\,\chi_s^2(\alpha_{\boldsymbol{m}})}{(2\pi\hbar)^{d-1}\,|\det \mathbb{D}_{\boldsymbol{m}}|}\, \delta(\tau T_H - T_{\boldsymbol{m}}) \ , \tag{7.82}$$

which can be readily evaluated using the sum rule for integrable skew products (6.73) which was derived in Sect. 6.5. Finally, we obtain

$$K_{\mathrm{diag}}(\tau) \sim \frac{(2s+1)\,|\Omega_E|}{(2\pi\hbar)^{d-1}\,T_H} = 1\,. \tag{7.83}$$

Thus for integrable systems with arbitrary spin s in the regime governed by the limits $T_H \to \infty$, $\tau \to 0$ and $\tau T_H \to \infty$, we find the same behaviour as the form factor of a Poisson process (7.42) shows for all τ. This observation confirms that, in the case of integrable classical dynamics, one should expect the quantum spectral statistics to be described by a Poisson process. Note however, that in order to obtain this result we had to extend the notion of integrability to the skew product of translational motion and spin precession. In the following section we will discuss what happens, if one tries to rely on integrability of the translational dynamics only.

7.3.4 Partially Integrable Systems

Having discussed the two extreme cases of strongly chaotic and completely integrable systems let us now consider an intermediate case. The translational dynamics ϕ_H^t shall be integrable but the restriction of the skew product flow Y_{sc}^t to an irrational (i.e. ergodic) Liouville–Arnold torus shall be ergodic on $\mathbb{T}^d \times S^2$ with respect to the measure $\mu_{\mathbb{T}^d} \times \mu_{S^2}$.

As in the previous cases we want to assume that degeneracies of contributions to the trace formula are due to symmetries like time reversal. Thus, we start from (7.68) with $g_{\mathcal{M}_s} = g$ for all tori. The value of g has to be determined for the particular system under consideration, for the following reason. Consider, for example, the time reversal invariant case, in which the image of a periodic orbit under time reversal can either be on the same torus or a different torus. We can not exclude the latter case as in the preceding section because, due to the non-integrability of Y_{sc}^t, we do not have explicit semiclassical quantisation conditions and thus no systematic degeneracy which would have to be removed before discussing spectral statistics.

From eqs. (7.68) and (3.117) we have

$$K_{\mathrm{diag}}(\tau) = \frac{g}{T_H} \sum_{\boldsymbol{m}} \frac{(2\pi)^{2d}\,(\overline{\chi_s}^{\boldsymbol{m}})^2}{(2\pi\hbar)^{d-1}\,|\det \mathbb{D}_{\boldsymbol{m}}|}\,\delta(\tau T_H - T_{\boldsymbol{m}})\,. \tag{7.84}$$

For a moment let us make the (wrong) assumption

$$(\overline{\chi_s}^{\boldsymbol{m}})^2 = \overline{\chi_s^2}^{\boldsymbol{m}}\,. \tag{7.85}$$

Then by comparing the sum rules (6.73) and (6.79) and employing the result of the preceding section, we would immediately find

$$K_{\mathrm{diag}}(\tau) \sim \frac{g}{2s+1}\,, \quad \tau \to 0\,. \tag{7.86}$$

Although the assumption (7.85) was not justified we will still find this asymptotics in a more sophisticated way.

To this end let us return to the derivation of sum rules for systems without spin (Sect. 6.3). We want to derive a variant of the result (6.44), namely a sum rule where the periodic orbit sum also includes an observable $a(\boldsymbol{I},\boldsymbol{\vartheta})$ averaged over the respective torus. This can be achieved, see e.g. [29], by replacing $\mathrm{Tr}\, U_{\boldsymbol{I}}^t$ by $\mathrm{Tr}(U_{\boldsymbol{I}}^t\, a(\boldsymbol{I},\cdot))$ in (6.35) and the following equations. One easily checks that now ergodicity of ϕ_H^t on a particular torus yields

$$\lim_{T\to\infty} \frac{1}{T}\int_0^T \mathrm{Tr}\left(U_{\boldsymbol{I}}^t\, a(\boldsymbol{I},\cdot)\right)\, \mathrm{d}t = \overline{a}^E\ , \tag{7.87}$$

with

$$\overline{a}^E := \mu_E(a) = \frac{1}{|\Omega_E|}\int_I\int_{\mathbb{T}^d} a(\boldsymbol{I},\boldsymbol{\vartheta})\,\delta(\overline{H}(\boldsymbol{I})-E)\,\mathrm{d}^d\vartheta\,\mathrm{d}^d I\ . \tag{7.88}$$

On the other hand, by performing exactly the same steps as from (6.38) to (6.44), we obtain the sum rule

$$\sum_{\boldsymbol{m}} \frac{(2\pi)^{2d}\,\overline{a}^{\boldsymbol{m}}}{|\det \mathbb{D}_{\boldsymbol{m}}|}\,\delta(T-T_{\boldsymbol{m}}) \sim |\Omega_E|\,\overline{a}^E\ , \tag{7.89}$$

with

$$\overline{a}^{\boldsymbol{m}} := \frac{1}{(2\pi)^d}\int_{\mathbb{T}^d} a(\boldsymbol{I}_{\boldsymbol{m}},\boldsymbol{\vartheta})\,\mathrm{d}^d\vartheta\ . \tag{7.90}$$

Now define the observable

$$a(\boldsymbol{I},\boldsymbol{\vartheta},t) := \frac{1}{(2\pi)^d}\int_{\mathbb{T}^d} \chi_s(\alpha(\boldsymbol{I},\boldsymbol{\vartheta},t))\,\chi_s(\alpha(\boldsymbol{I},\boldsymbol{\vartheta}',t))\,\mathrm{d}^d\vartheta'\ , \tag{7.91}$$

which depends on the additional parameter t. Thus also the average of a over a rational torus still depends on this parameter,

$$\overline{a}^{\boldsymbol{m}}(t) = \left[\frac{1}{(2\pi)^d}\int_{\mathbb{T}^d} \chi_s(\alpha(\boldsymbol{I}_{\boldsymbol{m}},\boldsymbol{\vartheta},t))\,\mathrm{d}^d\vartheta\right]^2\ , \tag{7.92}$$

and only for the choice $t = T_{\boldsymbol{m}}$ reduces to $\overline{a}^{\boldsymbol{m}}(T_{\boldsymbol{m}}) = (\overline{\chi_s}^{\boldsymbol{m}})^2$, i.e. the term which appears in the diagonal form factor (7.84). Leaving t arbitrary at first, we obtain the sum rule

$$\sum_{\boldsymbol{m}} \frac{(2\pi)^{2d}\,\overline{a}^{\boldsymbol{m}}(t)}{|\det \mathbb{D}_{\boldsymbol{m}}|}\,\delta(T-T_{\boldsymbol{m}}) \sim |\Omega_E|\,\overline{a}^E(t)\ , \tag{7.93}$$

with

$$\overline{a}^E(t) = \frac{(2\pi)^d}{|\Omega_E|}\int_I \delta(\overline{H}(\boldsymbol{I})-E)\left[\frac{1}{2\pi}\int_{\mathbb{T}^d} \chi_s(\alpha(\boldsymbol{I},\boldsymbol{\vartheta},t))\,\mathrm{d}^d\vartheta\right]^2\mathrm{d}^d I\ . \tag{7.94}$$

Now we can set $t = T$, which on the left hand side of the sum rule allows us to replace t by $T_{\boldsymbol{m}}$, i.e. we can use this sum rule in order to evaluate the diagonal form factor (7.84) in the limit $T_H \to \infty$, $\tau \to 0$ and $\tau T_H \to \infty$,

$$K_{\rm diag}(\tau) \sim \frac{g}{(2\pi\hbar)^{d-1}\,T_H}\,|\Omega_E|\,\overline{a}^E(\tau T_H) = \frac{g}{2s+1}\,\overline{a}^E(\tau T_H)\,. \qquad (7.95)$$

We are thus left with the task of determining the large-T asymptotics of $\overline{a}^E(T)$. So far we have only used ergodicity of the skew product on $\mathbb{T}^d \times S^2$, where $\mathbb{T}^d$ is a Liouville–Arnold torus with rationally independent frequencies. Now we have to impose the slightly stronger condition that correlations between $\alpha(\boldsymbol{I},\boldsymbol{\vartheta},T)$ and $\alpha(\boldsymbol{I},\boldsymbol{\vartheta}',T)$ decay for $\boldsymbol{\vartheta} \neq \boldsymbol{\vartheta}'$ and large T. In this case we have

$$\left[\frac{1}{2\pi}\int_{\mathbb{T}^d} \chi_s(\alpha(\boldsymbol{I},\boldsymbol{\vartheta},T))\,\mathrm{d}^d\vartheta\right]^2 \sim \frac{1}{2\pi}\int_{\mathbb{T}^d} \chi_s^2(\alpha(\boldsymbol{I},\boldsymbol{\vartheta},T))\,\mathrm{d}^d\vartheta\,, \quad T\to\infty\,. \qquad (7.96)$$

This condition is similar to (7.85), which we rejected. However, there it was for a rational torus, i.e. for a torus with periodic dynamics, whereas now we only need to impose it for typical, i.e. ergodic, tori. This makes sense if the field $\boldsymbol{\mathcal{B}}$ driving the spin dynamics is sufficiently random. According to (6.63) $\chi_s^2(\alpha) = I_s(\alpha)$, where in our case

$$I_s(\alpha(\boldsymbol{I},\boldsymbol{\vartheta},T)) = \int_{S^2} \delta_s(\boldsymbol{s}, R(\boldsymbol{I},\boldsymbol{\vartheta},T)\boldsymbol{s})\,\mathrm{d}\mu_{S^2}(\boldsymbol{s})\,, \qquad (7.97)$$

see (6.50), i.e.

$$\overline{a}^E(T) \sim \frac{(2\pi)^d}{|\Omega_E|}\int_{\mathcal{I}} \delta(\overline{H}(\boldsymbol{I})-E)\,\times$$
$$\times\,\underbrace{\frac{1}{(2\pi)^d}\int_{\mathbb{T}^d}\int_{S^2} \delta_s(\boldsymbol{s}, R(\boldsymbol{I},\boldsymbol{\vartheta},T)\boldsymbol{s})\,\mathrm{d}\mu_{S^2}(\boldsymbol{s})\,\mathrm{d}^d\vartheta}_{:=A(T)}\,\mathrm{d}^d I\,. \qquad (7.98)$$

If the skew product $Y_{\rm cl}^t$ restricted to $\mathbb{T}^d \times S^2$ was mixing, we would have

$$\lim_{T\to\infty} A(T) = 1\,. \qquad (7.99)$$

cf. (4.34). However, this is impossible, since the base flow on the torus, being given by irrational translations, can never be mixing. Nevertheless, it is still possible that correlations decay for pure spin observables, which is sufficient. Under these conditions we can indeed derive the asymptotics (7.86).

Summarising, we have obtained small-τ asymptotics for the spectral form factor which shows an intermediate behaviour between that of strongly chaotic systems, for which the form factor vanishes linearly, and integrable systems for which it tends to one. For partially integrable systems, which from the point of view of the classical dynamics show an intermediate behaviour, we obtain

$$0 < \lim_{\tau\to 0} K(\tau) < 1\,.^3 \qquad (7.100)$$

However, in order to obtain this result we needed a slightly stronger condition than just ergodicity of $Y_{\rm cl}^t$ for typical $\mathbb{T}^d \times S^2$. In the following section we will illustrate for a toy model that systems which show this behaviour indeed exist.

Recently intermediate statistics observed a lot of attention in the context of pseudointegrable or almost integrable dynamics, see e.g [37] and references therein. In particular it was conjectured that the so-called semi-Poisson statistics might serve as a characterisation for a new universality class. Bogomolny, Gerland and Schmit [35, 38] derived a random matrix model with short range interactions between neighbouring levels (in contrast to the Coulomb-like interaction in the Gaussian ensembles). Their model is reminiscent of (though not identical to) the situation discussed here, if we consider the spin-dependent term as a perturbation of an integrable system. Then the semiclassically small coupling between spin and translational degrees of freedom introduces an interaction between otherwise uncorrelated levels. The ensembles investigated in [35] also show the behaviour (7.100); in particular for the semi-Poisson case one has $K_{\rm sP}(\tau) \sim 1/2$, $\tau \to 0$.

7.4 Illustration: The $\boldsymbol{\sigma p}$-Rectangle

We discuss an example in order to illustrate our findings on the spectral statistics of partially integrable systems.

In Sect. 3.6.2 we introduced a class of model systems, two-dimensional billiards with a spin–orbit coupling term $\kappa\hat{\boldsymbol{s}}\hat{\boldsymbol{p}} = \kappa(\hbar/2)\mathrm{d}\pi_s(\boldsymbol{\sigma})\hat{\boldsymbol{p}}$ which we called $\boldsymbol{\sigma p}$-billiards. In Sects. 3.6.3 and 5.4.2 we have investigated the $\boldsymbol{\sigma p}$-torus, i.e. the Pauli equation on the rectangular domain

$$D = \left\{ \boldsymbol{x} \in \mathbb{R}^2 \,|\, 0 \le x_j \le a_j \right\} , \tag{7.101}$$

with periodic boundary conditions, which is an integrable system in the sense of Definition 1. To see this we have to consider the skew product $Y_{\rm cl}^t$ with classical Hamiltonian $H = \boldsymbol{p}^2$ (here and in the following we set $2m = 1$) and field $\boldsymbol{\mathcal{B}} = \kappa \boldsymbol{p}$ on $D \times S^2$. As a second skew product flow which commutes with $Y_{\rm cl}^t$ we can, for example, choose the flow with Hamiltonian $A = p_1^2$ and coupling field $\boldsymbol{\mathcal{B}}_A \propto \boldsymbol{p}$, see (5.43). The separability of the quantum mechanical equation of motion, the Pauli equation, can be destroyed by introducing Dirichlet boundary conditions, i.e. the requirement that the Pauli-spinor Ψ has to vanish on the boundary ∂D. For the classical system this implies elastic reflections on the walls which also destroys the integrability according to Definition 1: The skew product flow Y_A^t no longer commutes with $Y_{\rm cl}^t$ since

[3] The case $s = 0$ has to be excluded, because then there is no spin dynamics and the system is integrable; if we have $s = 1/2$ and $g = 2$, we obtain 1 in (7.86) but then we have to remove Kramers' degeneracy which forces the value down to 1/2.

at the boundary one component of $\boldsymbol{\mathcal{B}}_A$ changes its sign. Certainly, the translational flow ϕ_H^t is still integrable (which is guaranteed by $\{A, H\} = 0$), and thus we are potentially in the situation described in Sect. 7.3.4.

The spectral statistics of this billiard system with spin $\frac{1}{2}$ has been investigated numerically in [39]. When tuning the coupling κ the authors observe a transition from Poisson ($\kappa = 0$) to intermediate statistics, which could possibly be described by semi-Poisson statistics, and then for large κ back to Poisson statistics. We illustrate this behaviour in figures 7.2 and 7.1, where we show the level spacing density $P(s)$ and its distribution function $I(s)$.

The numerical data were obtained by expanding the components of the Pauli spinor Ψ in terms of the eigenfunctions of the corresponding spinless rectangular billiard with Dirichlet boundary conditions. Truncating this basis the Pauli Hamiltonian can be expressed as a (finite) matrix, which was diagonalised numerically. All plots shown in this section were calculated from approximately 400 eigenvalues. The ratio a_1/a_2 is chosen to $(\sqrt{5}-1)/2$. The level spacing density is defined by

$$\lim_{N\to\infty} \frac{1}{N} \#\{n \,|\, n \le N \;,\; a < x_{n+1} - x_n < b\} =: \int_a^b P(s)\,\mathrm{d}s \;, \qquad (7.102)$$

where x_n are the unfolded eigenvalues sorted by size, $x_1 \le x_2 \le \ldots$. Notice that we can realise the limit $N \to \infty$ by considering the energy levels in the interval $I(E, \hbar)$, as defined in (7.20), and taking the semiclassical limit. The distribution function, the integrated level spacing statistics $I(s)$, is given by

$$I(s) = \int_0^s P(s')\,\mathrm{d}s' \;. \qquad (7.103)$$

Before we can critically review the numerical results some remarks on the semiclassical limit are in order. When numerically investigating particular systems one does not usually vary $\hbar$ but fixes it to some value (as in nature its value is fixed) and has to investigate the system in a different but equivalent limit. For billiard systems (without spin) it can be shown, see e.g. [40], that $\hbar \to 0$, energy E fixed, corresponds to the limit of large momentum $p = \sqrt{E}$, i.e. $p \to \infty$ with $\hbar$ fixed. Similarly, the semiclassical limit can be connected to the high energy limit in other scaling systems, see e.g., [41] for a recent concise account on the relation of the semiclassical limit and the high energy limit. Here we only remark that the symbol of the Pauli Hamiltonian (3.126),

$$H_{\mathrm{P}}(\boldsymbol{p}, \boldsymbol{x}) = \boldsymbol{p}^2 + \frac{\kappa}{2}\mathrm{d}\pi_s(\boldsymbol{\sigma})\boldsymbol{p} \;, \quad (\hbar = 2m = 1) \;, \qquad (7.104)$$

viewed as a function of $p := |\boldsymbol{p}|$ in the limit $p \to \infty$ has a leading translational term (as in spinless billiards) and a subleading spin–orbit coupling term. Thus we conclude that in the high energy limit the semiclassical analysis presented so far applies to this system. So for any non-vanishing value of κ we expect $K(\tau) \sim g/(2s+1)$, $\tau \to 0$, according to (7.86). In particular

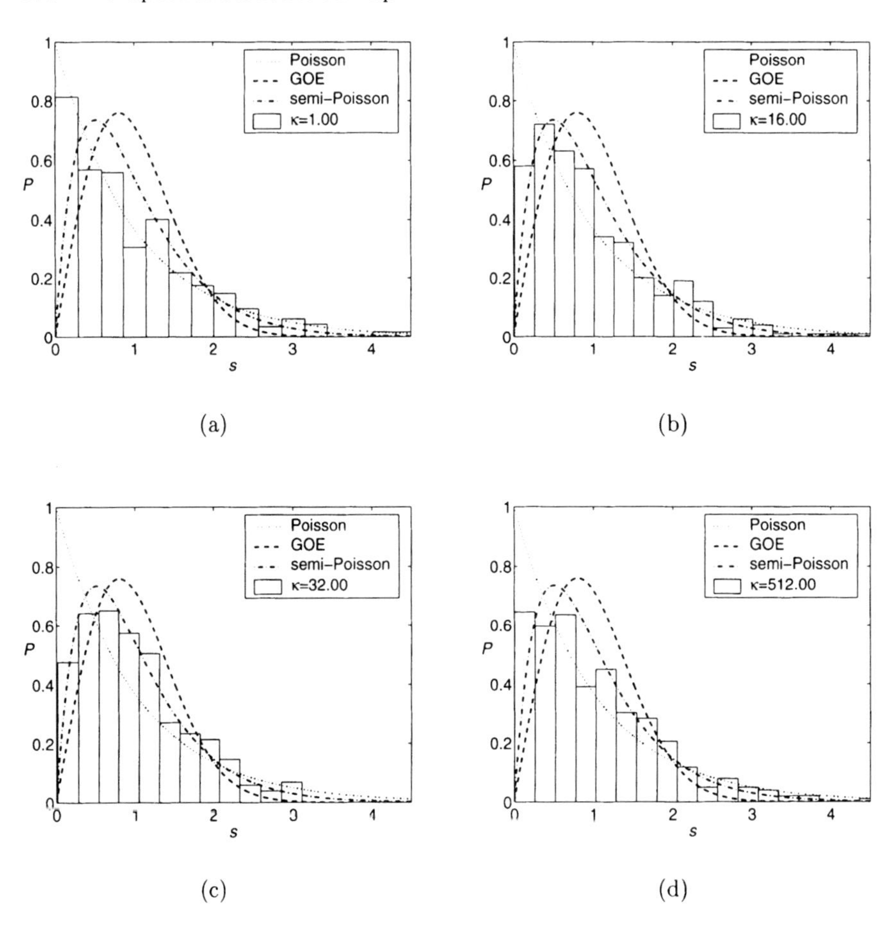

Fig. 7.1. The level spacing density $P(s)$ for a rectangular billiard with spin–orbit coupling. The coupling strength κ increases from (**a**)–(**d**).

for $s = 1/2$ and $g = 2$ (time reversal invariance) we have $K(\tau) \sim 1$, $\tau \to 0$ which, after removing Kramers' degeneracy, see (7.79), yields $\tilde{K} \sim 1/2$, which is consistent with semi-Poissonian statistics, cf. (7.43). For the remainder of the section we will drop the tilde on K again (as we do not put tildes on the analytical form factors with which we compare, anyway).

We can now understand the observations made in [39] on the basis of the semiclassical theory developed so far. When investigating the spectral statistics at both finite $\hbar$ and finite energy, which one has to do in numerical studies, it is the relative weight of the translational and spin–orbit terms in (3.126) which determines the statistics, which we should expect. For very small κ we can treat the spin–orbit term perturbatively and being close to an integrable system we expect Poissonian statistics, as in figures 7.1(a) and 7.2(a). By

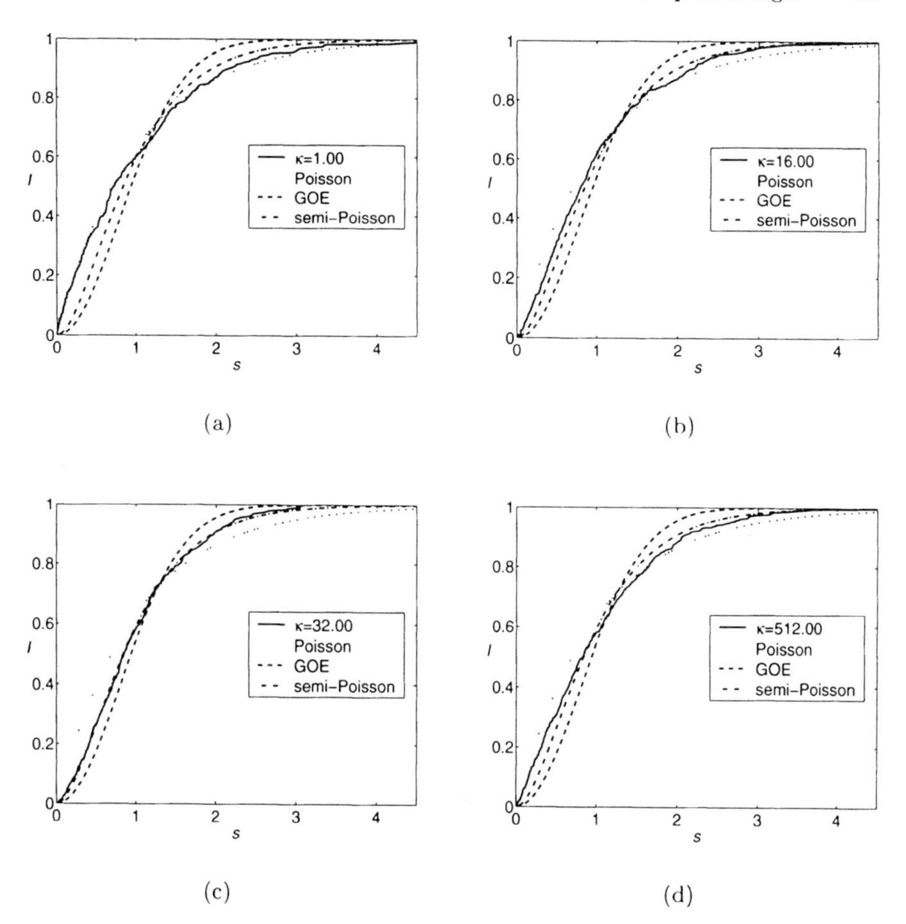

Fig. 7.2. The integrated level spacing $I(s)$ for a rectangular billiard with spin–orbit coupling. The coupling strength κ increases from (**a**)–(**d**). One observes a transition from Poisson to semi-Poison like statistics, (**a**)–(**c**), and back.

gradually increasing κ we observe a transition, cf. figures 7.1(b) and 7.2(b) to semi-Poisson like statistics, see figures 7.1(c) and 7.2(c). When going to even higher κ our semiclassical description breaks down. The subprincipal symbol $(\kappa/2)\mathrm{d}\pi_s(\boldsymbol{\sigma})\boldsymbol{p}$ is no longer small compared to the principal symbol and thus a semiclassical analysis which treats spin–orbit coupling as higher order term can no longer be applied. Instead one should now use the formalism outlined in Sect. 3.8, which leads to classical dynamics with polarised Hamiltonians

$$H_{m_s}(\boldsymbol{p},\boldsymbol{x}) = \boldsymbol{p}^2 + m_s\kappa|\boldsymbol{p}| \;, \quad m_s = -s,\ldots,s \;. \tag{7.105}$$

The flows $\phi^t_{H_{m_s}}$ are integrable since, for example, the observable p_1^2 commutes with the Hamiltonian and is also conserved during reflections at the boundary. Thus for large κ (with $\hbar$ fixed and E finite) we expect a transition back to Poissonian statistics, cf. figures 7.1(d) and 7.2(d).

We can also compare the numerical results directly to the statistics we calculated semiclassically in the preceeding section, the form factor $K(\tau)$. Being a distribution a plot of $K(\tau)$ itself does not make sense, since we would only observe seemingly erratic fluctuations, see Fig. 7.3. Moreover, its value at a single point ($\tau = 0$) which we are interested in, would not be well defined since we do not expect pointwise but only weak convergence. However, one can get an idea of the behaviour of the form factor by averaging over some interval $\Delta\tau$,

$$\frac{1}{\Delta\tau}\int_{\tau-\Delta\tau/2}^{\tau+\Delta\tau/2} K(\tau')\,\mathrm{d}\tau' \,, \tag{7.106}$$

i.e. we fold it with a test function as required, cf. eq. 7.33 (although for simplicity we choose a non-smooth window function). In Fig. 7.4 we show the numerically obtained averaged form factor together with the Poisson and semi-Poisson form factors, see eqs. (7.42) and (7.43). As discussed above we expect a transition from $K(\tau) \sim 1$, $\tau \to 0$ (Poisson), to $K(\tau) \sim 1/2$ (consistent with semi-Poisson) and back.

The form factors have been averaged over an interval of length $\Delta\tau = 0.2$. The data for τ-values below $\tau \approx 0.1$ is not shown, since it is less reliable depending strongly on how one takes into account the broadened singularity of $K(\tau)$ at $\tau = 0$; we comment on this problem in Appendix F. Therefore, to determine the asymptotics of K for small τ from the plots in Fig. 7.4, we have to extrapolate from outside this region to $\tau = 0$. Then in Fig. 7.4 we observe the expected transition simultaneously with the transition observed for the level spacings, see Figs. 7.1 and 7.2. Furthermore, within the numerical

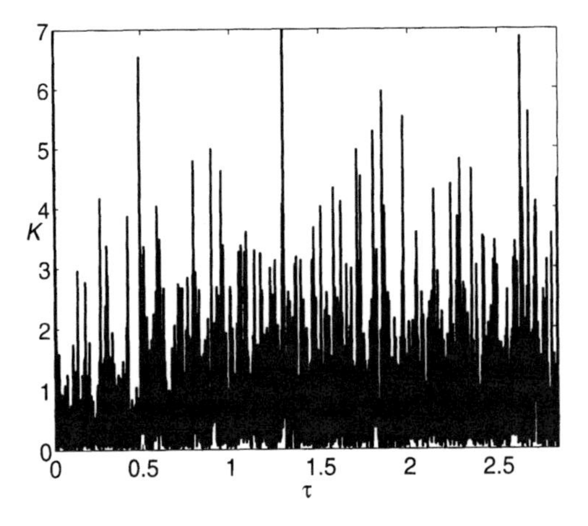

Fig. 7.3. The form factor $K(\tau)$ for the rectangular $\boldsymbol{\sigma p}$-billiard with spin $s = 1/2$ and coupling strength $\kappa = 32$.

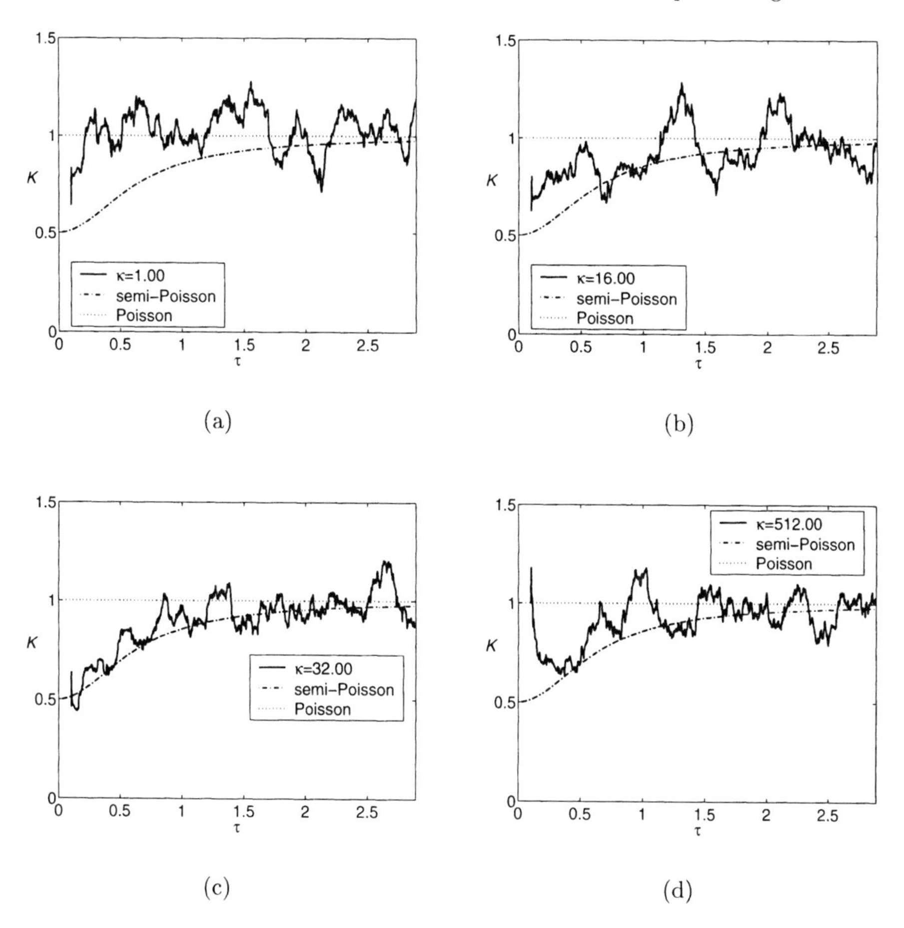

Fig. 7.4. The averaged spectral form factor ($\Delta\tau = 0.2$) for a rectangular billiard with spin–orbit coupling. The coupling strength κ increases from (**a**)–(**d**) and the transition observed is simultaneous to that for the distribution of level spacings, see figures 7.1 and 7.2.

accuracy, we also conclude that the semi-Poisson form factor (7.43) provides a reasonable fit to our data.

In order to further test our result (7.86) we can now vary the spin s. Before we do so a remark on the choice of the pairs of parameters of spin s and coupling strength κ is in order. If we want to determine the scale for the transition from Poisson ($\kappa = 0$) to intermediate statistics ($\kappa \neq 0$) we should ask "how far away" the classical spin vector is allowed to precess from its initial position. If this distance is not large we cannot assume uncorrelated spin rotation angles in (7.96). When looking at a the value of the form factor for a given τ the relevant time scale is set by τT_H. The momentary frequency of spin precession,

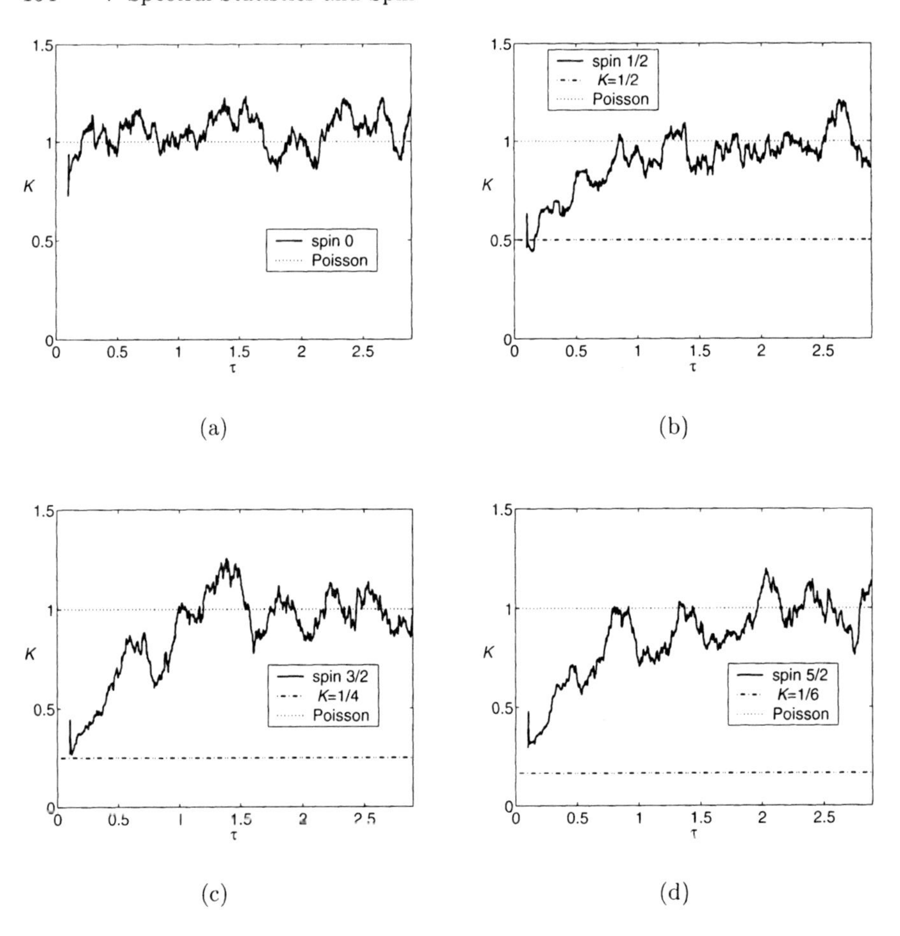

Fig. 7.5. The averaged spectral form factor ($\Delta\tau = 0.2$) for a rectangular billiard with spin–orbit coupling. The asymptotic value expected for $\tau \to 0$ according to (7.86) is indicated as horizontal dash-dotted line. The spin s increases from (a)–(d); simultaneously the coupling strength decreases according to $\kappa(2s+1) = const$, from $\kappa = 32$ (b) to $\kappa = 10\frac{2}{3}$ (d).

on the other hand, is given by $|\boldsymbol{\mathcal{B}}| = \kappa|\boldsymbol{p}|$. It is the product $\tau T_H|\boldsymbol{\mathcal{B}}|$ which has to be large in order to observe intermediate statistics. Notice that asymptotically we have $\tau T_H \to \infty$ and thus expect an abrupt transition from Poisson to intermediate statistics at $\kappa = 0$. For the finite energy range which is accessible numerically we can obtain the following scaling argument from the above discussion. Since, through the mean density $\bar{d}$, the Heisenberg time depends on the spin as $T_H \propto (2s+1)$, we have to keep the product $(2s+1)\kappa$ constant when varying spin and coupling strength in order to obtain comparable results.

In Fig. (7.5) we show the form factor for the rectangular $\boldsymbol{\sigma p}$-billiard with spin $2s = 0, 1, 3, 5$. For $s = 0$ we expect $K \sim 1$, $\tau \to 0$ (Poisson), whereas for half-integer s we expect $K \sim 1/(2s+1)$, $\tau \to 0$ (after removing Kramers' degeneracy). It appears that this prediction works reasonably well for spin $s = 1/2$ and $s = 3/2$, figures 7.5(b) and 7.5(c), although a definitive statement would require more extensive numerical studies. In the case of spin $s = 5/2$ one can hardly tell how the curve has to be extrapolated to $\tau = 0$. Therefore, we cannot draw a final conclusion on whether our prediction can be verified in this case.

To further illustrate these problems, in Fig. 7.6 we show again the numerically obtained form factor (spin $s = 5/2$, $\kappa = 10\frac{2}{3}$) together with the form factors for Poisson (dotted), semi-Poisson (dash-dotted), and GOE (dashed) statistics. Within the numerical accuracy of the present data $K(\tau)$ could even follow GOE-statistics for small τ (dropping down to zero at $\tau = 0$) although it may well behave as predicted by (7.86).

Summarising, we have shown that there exist systems for which the translational dynamics is integrable whereas the combined spin and translational dynamics (the skew product Y^t_{cl}) is not, which was the main goal of this section. Moreover, we have shown in a preliminary numerical discussion that the semiclassical theory developed in Sect. 7.3.4 can be useful for describing the phenomena arising in this context. Since the focus of the present work is on analytical, semiclassical techniques we will stop at this point and postpone a final quantitative analysis of the validity of formula (7.86) until more extensive numerical studies have been carried out.

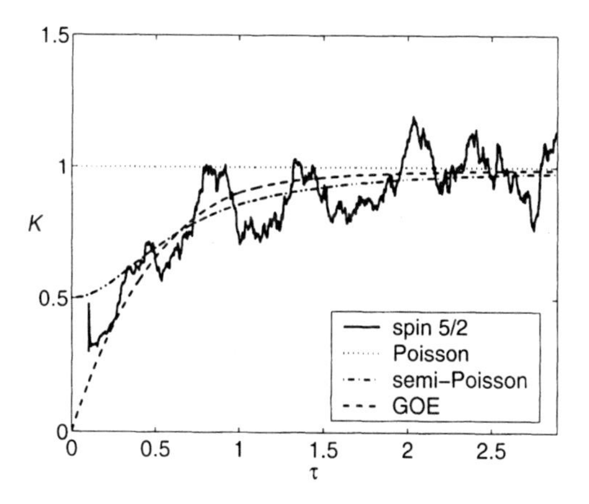

Fig. 7.6. The form factor $K(\tau)$ for the rectangular $\boldsymbol{\sigma p}$-billiard with spin $s = 5/2$ and coupling strength $\kappa = 10\frac{2}{3}$.

7.5 Other Statistical Measures

In this section we give a brief overview of how the methods developed so far can be used in order to describe other statistical measures besides the two-point form factor. We state the results for the different situations, and comment on how these methods can be extended in order to gain further information.

7.5.1 The Number Variance

The first statistic we discuss is the number variance $\Sigma^2(L)$. Denote by $\mathcal{N}(x)$ the integrated spectral density or spectral staircase function of the unfolded spectrum,

$$\mathcal{N}(x) := \int^{x} \mathscr{d}(x')\, \mathrm{d}x' \; . \tag{7.107}$$

If we ask how much the number of (unfolded) levels in an interval of length L differs from its mean, which is also given by L, this can be measured by the number variance

$$\Sigma^2(L) := \left\langle (\mathcal{N}(x'+L) - \mathcal{N}(x') - L)^2 \right\rangle_{x'} \; . \tag{7.108}$$

This quantity is readily expressed in terms of the original (non-unfolded) spectrum by

$$\Sigma^2(L) = \left\langle \left(N\left(E' + \frac{L}{\bar{d}}\right) - N(E') - L \right)^2 \right\rangle_{E'} \; , \tag{7.109}$$

where the integrated density of states is defined by

$$N(E) = \int^{E} d(E')\, \mathrm{d}E' \; . \tag{7.110}$$

Let us first investigate how the number variance behaves for small arguments L. This behaviour is independent of the particular system but depends only on the spectrum being discrete and non-degenerate [34]. For small L we either find one eigenvalue between E' and $E' + L/\bar{d}$ or no eigenvalue at all. The probability of finding one eigenvalue is given by L (the mean density $\bar{d}$ multiplied by the length $L/\bar{d}$ of the interval) and, by normalisation, the probability of finding no eigenvalue is given by $1 - L$. Thus we have

$$\Sigma^2(L) = L(1-L)^2 + (1-L)L^2 = L + \mathcal{O}\left(L^2\right) \; . \tag{7.111}$$

For large L we expect to see the classical characteristics of the system under consideration. For integrable classical dynamics we compare to a Poisson process for which

$$\Sigma^2_{\text{Poisson}}(L) = L \tag{7.112}$$

also for large L. The random matrix ensembles, to which we have to compare in the chaotic case, show only logarithmic growth for large L [7],

$$\Sigma^2(L) \sim \begin{cases} \frac{1}{\pi^2} \log L \;, & \text{GUE} \\ \frac{2}{\pi^2} \log L \;, & \text{GOE} \\ \frac{1}{2\pi^2} \log L \;, & \text{GSE} \end{cases} \;, \quad L \to \infty \tag{7.113}$$

For semi-Poisson statistics, which we have seen also has some relevance for systems with spin, one has

$$\Sigma^2_{\text{sP}}(L) = \frac{L}{2} + \frac{1}{8}\left(1 - \mathrm{e}^{-4L}\right) \sim \frac{L}{2} \;, \quad L \to \infty \;. \tag{7.114}$$

However, when one numerically analyses the number variance of a quantum system one does not observe any of these asymptotics for large L, but instead $\Sigma^2(L)$ saturates, and for L greater than some value which we will call $L_{\max}$ shows irregular oscillations about some mean value Σ^2_∞. We will see that semiclassically one can understand both the saturation and the universal behaviour for L large but smaller than $L_{\max}$.

Like the two-point correlation function $R_2(s)$ and the form factor $K(\tau)$ the number variance measures two-point correlations. This can be seen from eq. (2.52) where the number variance is expressed as an integral transform of the spectral form factor. Another popular statistic also measuring two-point correlations is the spectral rigidity, see [7] for a definition. Qualitatively the spectral rigidity and the number variance show the same behaviour, both exhibiting a universality and a saturation regime as described above. The first one to explain this on a semiclassical level was Berry [17]. He investigated the spectral rigidity for which one also has a representation as an integral transform of the form factor. Into this formula Berry inserted a model form factor obtained from the semiclassical asymptotics for small τ and the behaviour (7.41) for large τ, between which he interpolated in different ways. He then showed that the relevant features of the spectral rigidity, in particular the saturation, are independent of the interpolation. Similar approaches to the number variance can, e.g., be found in [13, 34]. In a slight variation we propose to directly extract the relevant information from a semiclassical representation of the number variance. But in spirit our treatment closely follows [17].

In order to be able to use the trace formula we introduce a smoothed staircase function

$$N_\varepsilon(E') := \int^{E'} d_\varepsilon(E'')\,\mathrm{d}E'' \;. \tag{7.115}$$

Working in the same spectral interval $I(E,\hbar)$ as described in Sect. 7.3, for the mean part of the integrand we can simply substitute the constant $\overline{d} \equiv \overline{d}(E)$. If there are no bifurcations the asymptotics of the integral over the periodic

orbit sum can be found by integrating only the rapidly oscillating exponential as follows.
CLAIM:

$$N_\varepsilon^{\mathrm{fl}}(E') \sim \sum_{\mathcal{M}_s} \frac{\hat{\eta}(\varepsilon T_{\mathcal{M}_s})}{2\pi \mathrm{i}\, T_{\mathcal{M}_s}} \mathcal{A}_{\mathcal{M}_s} \,\mathrm{e}^{(\mathrm{i}/\hbar) S_{\mathcal{M}_s}} , \tag{7.116}$$

in the sense that

$$\frac{\mathrm{d}}{\mathrm{d}E'} \left(\sum_{\mathcal{M}_s} \frac{\hat{\eta}(\varepsilon T_{\mathcal{M}_s})}{2\pi \mathrm{i}\, T_{\mathcal{M}_s}} \mathcal{A}_{\mathcal{M}_s} \,\mathrm{e}^{(\mathrm{i}/\hbar) S_{\mathcal{M}_s}} \right) = d_\epsilon^{\mathrm{fl}}(E') \left(1 + \mathcal{O}(\hbar)\right) . \tag{7.117}$$

Proof. If the (families of) orbits $\mathcal{M}_s$ at different energies can be smoothly related (which is possible since we have excluded bifurcations), we may interchange the E'-differentiation and the periodic orbit sum. In general $S_{\mathcal{M}_s}$, $T_{\mathcal{M}_s}$ and $\mathcal{A}_{\mathcal{M}_s}$ depend on E', and we have the relation $T_{\mathcal{M}_s} = \mathrm{d}S_{\mathcal{M}_s}/\mathrm{d}E'$. Then it follows that

$$\begin{aligned} \frac{\mathrm{d}}{\mathrm{d}E'} N_\varepsilon^{\mathrm{fl}}(E') &= \sum_{\mathcal{M}_s} \mathrm{e}^{(\mathrm{i}/\hbar) S_{\mathcal{M}_s}} \left(\frac{\hat{\eta}(\varepsilon T_{\mathcal{M}_s})}{2\pi\hbar} \mathcal{A}_{\mathcal{M}_s} + \frac{\mathrm{d}}{\mathrm{d}E'} \frac{\hat{\eta}(\varepsilon T_{\mathcal{M}_s})}{2\pi \mathrm{i}\, T_{\mathcal{M}_s}} \mathcal{A}_{\mathcal{M}_s} \right) \\ &= \sum_{\mathcal{M}_s} \frac{\hat{\eta}(\varepsilon T_{\mathcal{M}_s})}{2\pi\hbar} \mathcal{A}_{\mathcal{M}_s} \,\mathrm{e}^{(\mathrm{i}/\hbar) S_{\mathcal{M}_s}} \\ &\qquad \times \Bigg(1 + \underbrace{\frac{2\pi\hbar}{\hat{\eta}(\varepsilon T_{\mathcal{M}_s}) \mathcal{A}_{\mathcal{M}_s}} \frac{\mathrm{d}}{\mathrm{d}E'} \frac{\hat{\eta}(\varepsilon T_{\mathcal{M}_s})}{2\pi \mathrm{i}\, T_{\mathcal{M}_s}} \mathcal{A}_{\mathcal{M}_s}}_{=\mathcal{O}(\hbar)} \Bigg) . \end{aligned} \tag{7.118}$$

□

Thus, the smoothed spectral staircase is given by

$$N_\varepsilon(E') \sim \overline{d} E' + N_\varepsilon^{\mathrm{fl}}(E') , \tag{7.119}$$

and the smoothed number variance can be defined as

$$\Sigma_\varepsilon^2(L) := \left\langle \left(N_\varepsilon^{\mathrm{fl}}\left(E' + \frac{L}{\overline{d}}\right) - N_\varepsilon^{\mathrm{fl}}(E') \right)^2 \right\rangle_{E'} . \tag{7.120}$$

We are now interested in the large-L asymptotics in the semiclassical limit. The latter implies $\overline{d} \to \infty$, and in order to properly stay in the interval $I(E, \hbar)$ we have to ensure that $L/\overline{d}$ tends to zero. Thus, we have to examine (7.120) in the combined limit

$$\overline{d} \to \infty , \quad L \to \infty \quad \text{and} \quad L/\overline{d} \to 0 . \tag{7.121}$$

To this end we proceed as in Sect. 7.3.1. Expanding the actions as in (7.59) we obtain

$$\Sigma_\varepsilon^2(L) = \left\langle \left(\sum_{\mathcal{M}_s} \frac{\hat{\eta}(\varepsilon T_{\mathcal{M}_s})}{2\pi \mathrm{i}\, T_{\mathcal{M}_s}} \mathcal{A}_{\mathcal{M}_s} \left(\exp\left(\mathrm{i} \frac{T_{\mathcal{M}_s} L}{\hbar \bar{d}} \right) - 1 \right) \right. \right.$$
$$\left. \left. \times \exp\left(\frac{\mathrm{i}}{\hbar} S_{\mathcal{M}_s} + \frac{\mathrm{i}}{\hbar} T_{\mathcal{M}_s}(E'-E) \right) \right)^2 \right\rangle_{E'} , \tag{7.122}$$

where we have suppressed the energy argument E of $S_{\mathcal{M}_s}$, $T_{\mathcal{M}_s}$ and $\mathcal{A}_{\mathcal{M}_s}$. Following the same arguments as in Sect. (7.3.1) we only keep the diagonal terms when performing the E'-average, which leaves us with the diagonal number variance

$$\begin{aligned} \Sigma_{\varepsilon,\mathrm{diag}}^2(L) :&= \frac{g}{2\pi^2} \sum_{\mathcal{M}_s} \hat{\eta}(\varepsilon T_{\mathcal{M}_s})\, \hat{\eta}(-\varepsilon T_{\mathcal{M}_s})\, \frac{|\mathcal{A}_{\mathcal{M}_s}|^2}{T_{\mathcal{M}_s}^2} \left(1 - \cos\left(\frac{T_{\mathcal{M}_s} L}{\hbar \bar{d}} \right) \right) \\ &= \frac{g}{\pi^2} \sum_{\mathcal{M}_s, T_{\mathcal{M}_s}>0} \hat{\eta}(\varepsilon T_{\mathcal{M}_s})\, \hat{\eta}(-\varepsilon T_{\mathcal{M}_s})\, \frac{|\mathcal{A}_{\mathcal{M}_s}|^2}{T_{\mathcal{M}_s}^2} \left(1 - \cos\left(\frac{T_{\mathcal{M}_s} L}{\hbar \bar{d}} \right) \right) , \end{aligned} \tag{7.123}$$

where the notation is exactly as in (7.68). On the second line we have exploited that the expression under the sum is symmetric with respect to $T_{\mathcal{M}_s} \to -T_{\mathcal{M}_s}$. The remaining periodic orbit sums can be evaluated using the sum rules of Sect. 6.5. We demonstrate this for strongly chaotic systems and state the results for the other cases discussed in Sects. (7.3.3) and (7.100) which are obtained analogously. For strongly chaotic systems we have (see (7.73) and also Sect. 6.5)

$$\sum_{\mathcal{M}_s} |\mathcal{A}_{\mathcal{M}_s}|^2\, \delta(T - T_{\mathcal{M}_s}) \sim T . \tag{7.124}$$

Upon multiplication with

$$\hat{\eta}(\varepsilon T)\, \hat{\eta}(-\varepsilon T)\, \frac{1 - \cos\left(TL/(\hbar \bar{d})\right)}{T^2} \tag{7.125}$$

and integration over T we obtain

$$\Sigma_{\varepsilon,\mathrm{diag}}^2(L) \sim \frac{g}{\pi^2} \int_0^{T^*} \hat{\eta}(\varepsilon T)\, \hat{\eta}(-\varepsilon T)\, \frac{1 - \cos\left(TL/(\hbar \bar{d})\right)}{T}\, \mathrm{d}T , \tag{7.126}$$

where T^* is the cut-off provided by the compact support of $\hat{\eta}$. Obviously this expression diverges if we take the limit $\varepsilon \to 0$ and thus remove the cut-off. Therefore, we have to come back to our rule formulated in Sect. 7.3.1 of only gradually removing the cut-off in the semiclassical limit, such that we are just able to resolve correlations on the scale of the mean level spacing. This procedure implies $T^* = \mathcal{O}(T_H)$ and by the asymptotic expansion of the cosine integral we find

$$\Sigma_{\mathrm{diag}}^2(L) \sim \frac{g}{\pi^2} \log L , \quad L \to \infty , \tag{7.127}$$

which for $g = 1$ and $g = 2$ agrees with the GUE- and GOE-asymptotics (7.113), respectively. In the case of time reversal invariant systems with half-integer spin we have to take into account Kramers' degeneracy as described in Sect. 7.3.2. By making the substitutions

$$N(E') \to \tilde{N}(E') = \frac{1}{2}N(E') \quad \text{and} \quad \overline{d} \to \overline{\tilde{d}} = \frac{1}{2}\overline{d} \tag{7.128}$$

in (7.109) we find the relation

$$\tilde{\Sigma}^2(L) = \frac{1}{4}\Sigma^2(2L) \tag{7.129}$$

between the number variance $\Sigma^2(L)$ of the degenerate spectrum and the number variance $\tilde{\Sigma}^2(L)$ of the spectrum with Kramers' degeneracy removed. For time reversal invariant systems with half-integer spin we finally obtain

$$\tilde{\Sigma}^2_{\text{diag}} \sim \frac{1}{2\pi^2} \log L , \tag{7.130}$$

which agrees with the GSE-asymptotics (7.113).

By analogous considerations for integrable systems one finds

$$\Sigma^2_{\text{diag}}(L) \sim L , \quad L \to \infty , \tag{7.131}$$

which is in agreement with the number variance for a Poisson process, whereas for partially integrable systems we obtain

$$\Sigma^2_{\text{diag}}(L) \sim \frac{g}{2s+1} L , \quad L \to \infty . \tag{7.132}$$

However, when numerically looking at the number variance of a given system, one usually observes a different behaviour. The number variance first follows the predictions from random matrix theory or the Poisson process, respectively, but for large L one finds that $\Sigma^2(L)$ stops growing and only shows irregular oscillations about some saturation value Σ^2_∞. This behaviour can also be explained using the trace formula [17]. To this end revisit the derivation of the above results. In the step from (7.120) to (7.122) we expanded the actions $S_{\mathcal{M}_s}$ about the energy E. This only makes sense if the distance between E' and $E' + L/\overline{d}$ is much smaller than the length of a typical oscillation of $N^{\text{fl}}_\varepsilon$ which according to (7.122) is given by $2\pi\hbar/T_{\mathcal{M}_s}$. If in contrast

$$\frac{L}{\overline{d}} \gg \frac{2\pi\hbar}{T_{\mathcal{M}_s}} , \tag{7.133}$$

then $N^{\text{fl}}_\varepsilon(E')$ and $N^{\text{fl}}_\varepsilon(E' + L/\overline{d})$ should be treated as uncorrelated. In that sense the typical periods $T_{\mathcal{M}_s}$ of short periodic orbits define a classical time scale T_{cl} of the system which in turn determines the maximal argument

$$L_{\text{max}} := T_H/T_{\text{cl}} \tag{7.134}$$

of the number variance, where the universality regime ends, i.e. above which we do not expect correlations according to, e.g, random matrix theory. Notice that this regime is still within the range of the combined limit (7.121), since we can simultaneously have

$$L/\overline{d} \to 0 \quad \text{but} \quad L/T_H \to \infty\,. \tag{7.135}$$

In the saturation regime the number variance can be estimated as

$$\begin{aligned}\Sigma_\varepsilon^2(L) &\approx \left\langle \left[N_\varepsilon^{\mathrm{fl}}\left(E'+\frac{L}{\overline{d}}\right)\right]^2\right\rangle_{E'} + \left\langle \left[N_\varepsilon^{\mathrm{fl}}(E')\right]^2\right\rangle_{E'} \\ &\approx 2\left\langle \left[N_\varepsilon^{\mathrm{fl}}(E')\right]^2\right\rangle_{E'}\,.\end{aligned} \tag{7.136}$$

Inserting the trace formula, performing the diagonal approximation and evaluating the remaining periodic orbit sum with a classical sum rule, as we did for (7.123), one obtains for strongly chaotic systems [13, 17],

$$\Sigma_{\varepsilon,\mathrm{diag}}^2(L) \sim \frac{g}{2\pi^2}\log T_H =: \Sigma_\infty^2\,, \tag{7.137}$$

and similarly for integrable and partially integrable systems.

In the semiclassical limit both L_{max} and the saturation value Σ_∞^2 go to infinity and thus universality is restored on arbitrarily long scales in L. In this sense saturation is related to the finite length and resolution of a sequence of levels with which one has to deal in numerical experiments. In the introductory Chap. 2 on oscillators we have simulated this behaviour by truncating the periodic orbit sum and thus restricting the resolution in energy.

Notice that on the basis of the crude approximations involved in the above argument neither the value L_{max} at which saturation sets in nor the saturation value Σ_∞^2 can be predicted explicitly. Only their scaling with $\hbar$ or energy is accessible within semiclassical considerations. This behaviour has been verified numerically, see e.g. [34, 42, 43].

7.5.2 The Value Distribution of $N_\varepsilon^{\mathrm{fl}}$

As another measure for spectral correlations and thus for quantum chaos Aurich, Bolte and Steiner proposed to investigate the value distribution of $N_\varepsilon^{\mathrm{fl}}$ [44], see also [45]. Since this quantity describes the fluctuations of the smoothed spectral staircase function N_ε about its mean $\overline{N}$ the mean value of $N_\varepsilon^{\mathrm{fl}}$ vanishes. According to the discussion of the number variance $\Sigma^2(L)$ in the preceeding section the variance of $N_\varepsilon^{\mathrm{fl}}$ is given by $\Sigma_\infty^2/2$. In the semiclassical limit the saturation value Σ_∞^2 of the number variance diverges, i.e. the fluctuations grow larger and larger, and one should suitably normalise $N_\varepsilon^{\mathrm{fl}}$ before investigating its value distribution. To this end one defines the quantity

$$W_\varepsilon(E') := \frac{N_\varepsilon^{\mathrm{fl}}(E')}{\sqrt{\frac{1}{2}\Sigma_\infty^2}} . \tag{7.138}$$

It is conjectured that the value distribution of $W_\varepsilon(E')$ assumes a limit distribution in the semiclassical limit, i.e.

$$\lim_{\hbar\to 0} \langle \delta(W_\varepsilon(E') - w)\rangle_{E'} = \int_{\mathbb{R}} p(w)\,\mathrm{d}w , \tag{7.139}$$

where the limit $\varepsilon \to 0$ is taken simultaneously with the semiclassical limit $\hbar \to 0$ such that the effective cut-off of the periodic orbit sum for $N_\varepsilon^{\mathrm{fl}}$ is of the order of the Heisenberg time T_H, cf. the discussion in Sects. 7.3.1 and 7.5.1. Due to our previous remark $p(w)$ must have zero mean and unit variance. Aurich, Bolte and Steiner conjecture that for strongly chaotic systems (with only isolated periodic orbits) the limit distribution is given by a Gaussian,

$$p(w) = \frac{1}{\sqrt{2\pi}}\,\mathrm{e}^{-w^2/2} , \tag{7.140}$$

whereas for integrable systems they expect a non-universal but non-Gaussian limit distribution.

There is strong numerical evidence in favour of this conjecture, see e.g. [44, 46, 47] and references therein. A semiclassical argument based on the Gutzwiller trace formula was given in [13]. We refrain from repeating any details here but only remark that the semiclassical argument is of a rather general nature and should not be substantially modified by the spin contribution. Thus, in the following, we will assume the conjecture to hold for all strongly chaotic systems independent of their spin, provided that the translational dynamics has only isolated periodic orbits and that the skew product flow Y_{cl}^t is ergodic.

7.5.3 $R_2(s)$ and the Bogomolny-Keating Bootstrap

For the semiclassical analysis of some spectral functions (the number variance $\Sigma^2(L)$ and the value distribution $p(w)$ of $N_\varepsilon^{\mathrm{fl}}$) it was crucial to cut off the periodic orbit sums at the order of the Heisenberg time T_H. By that means all expressions appearing in the discussion are smoothed, which was our intention in the first place. However, one could argue that by smoothing the spectral density one somehow forgets that the spectrum is discrete and that this might affect the results in an unwanted way. On the other hand it is known that there exist methods for extracting a discrete spectrum of semiclassical eigenvalues from smoothed spectral functions, see e.g. [48–51]. These semiclassical eigenvalues are observed to be good approximations to the real spectrum, see e.g. [52, 53], and thus Bogomolny and Keating [24] proposed to analytically discuss the correlations of the approximate spectrum. This is a bootstrap method in the sense that technically one uses a truncated trace formula giving rise to a non-singular density of states but still discusses the correlations for a discrete spectrum. The procedure is most conveniently carried out for the two-point correlation function $R_2(s)$.

We will see that in this way one can obtain the leading oscillatory and the leading non-oscillatory contributions to the large-s asymptotics of $R_2(s)$. By general properties of the Fourier transformation non-exponentially decaying terms are related to singularities of the Fourier transform, here the spectral form factor $K(\tau)$, or its derivatives on the real axis. In this sense the result which we will find for the non-oscillatory term in the large-s expansion of $R_2(s)$ corresponds to the diagonal approximation of the form factor, which in a (real) vicinity of $\tau = 0$ behaves like $K(\tau) \approx \alpha|\tau|$ with some non-zero slope α, i.e. the form factor has a kink at $\tau = 0$. The leading oscillatory term of $R_2(s)$ will then provide additional information which goes beyond the diagonal approximation. In the process of the calculations we will use a variant of the conjecture discussed in Sect. 7.5.2 and thus the following discussion applies to strongly chaotic systems only.

The two-point correlation functions of the Gaussian ensembles of random matrix theory and their large-s asymptotics are given by [7]

$$R_2^{\mathrm{GUE}}(s) = \delta(s) - \left(\frac{\sin(\pi s)}{\pi s}\right)^2 = \delta(s) - \frac{1}{2(\pi s)^2} + \frac{\cos(2\pi s)}{2(\pi s)^2} , \tag{7.141}$$

$$\begin{aligned} R_2^{\mathrm{GOE}}(s) &= \delta(s) - \int_s^\infty \frac{\sin(\pi s')}{\pi s'}\,\mathrm{d}s' \left(\frac{\mathrm{d}}{\mathrm{d}s}\frac{\sin(\pi s)}{\pi s}\right) - \left(\frac{\sin(\pi s)}{\pi s}\right)^2 \\ &\sim -\frac{1}{(\pi s)^2} + \frac{3+\cos(2\pi s)}{2(\pi s)^4} , \qquad s \to \infty , \end{aligned} \tag{7.142}$$

and

$$\begin{aligned} R_2^{\mathrm{GSE}}(s) &= \delta(s) - \left(\frac{\sin(2\pi s)}{2\pi s}\right)^2 + \int_0^s \left(\frac{\sin(2\pi s')}{2\pi s'}\right) \mathrm{d}s' \left(\frac{\mathrm{d}}{\mathrm{d}s}\frac{\sin(2\pi s)}{2\pi s}\right) \\ &\sim \frac{\pi}{2}\frac{\cos(2\pi s)}{2\pi s} - \frac{1+\frac{\pi}{2}\sin(2\pi s)}{(2\pi s)^2} , \qquad s \to \infty , \end{aligned} \tag{7.143}$$

respectively.[4]

The Bogomolny-Keating bootstrap relies on the following quantisation condition introduced in [51, 54],

$$N_\varepsilon(E_n(T)) \overset{!}{=} n + \frac{1}{2} , \qquad n \in \mathbb{N}_0 , \tag{7.144}$$

where we explicitly indicated the dependence of the semiclassical eigenvalues on the effective cut-off time T which is introduced by the test function $\hat{\eta}(\varepsilon T_\gamma)$,

[4] The function $R_2(s)$ in our notation is related to the function $Y_2(r)$ in [7] by $R_2(s) = \delta(s) - Y_2(s)$.

cf. (7.116). The quantisation rule (7.144) is based on the observation that, although the truncated staircase function cannot reproduce the sharp steps of $N(E)$, a good approximation for the energies at which the steps occur can be obtained by searching for the positions where $N_\varepsilon(E)$ lies exactly between two plateaus of height n and $n+1$. Since in the case of time reversal invariant systems with half-integer spin, due to Kramers' degeneracy, the typical step size is no longer one but two, eq. (7.144) should be changed to

$$N_\varepsilon(E_n(T)) \stackrel{!}{=} 2n+1\,, \quad n \in \mathbb{N}_0\,. \tag{7.145}$$

The approximate spectrum $\{E_n(T)\}$ then already has Kramers' degeneracy removed. In the following we will write

$$N_\varepsilon(E_n(T)) \stackrel{!}{=} \chi\left(n+\frac{1}{2}\right)\,, \quad n \in \mathbb{N}_0\,, \tag{7.146}$$

to account for the cases with ($\chi = 2$) and without ($\chi = 1$) Kramers' degeneracy, respectively.

Let us introduce the spectral density of the approximate spectrum by

$$\begin{aligned} D_T(E) :&= \sum_{n\in\mathbb{N}_0} \delta(E - E_n(T)) \\ &= \sum_{n\in\mathbb{N}_0} |d_\varepsilon(E)|\,\delta\Big(N_\varepsilon(E) - \chi\left(n+\frac{1}{2}\right)\Big)\,. \end{aligned} \tag{7.147}$$

If in (7.116) we choose positive test functions, more precisely $\eta(\omega) \geq 0\ \forall\ \omega \in \mathbb{R}$, then $d_\varepsilon(E)$ is also positive and we can omit the modulus in (7.147). By Poisson summation (see Appendix A) we obtain

$$\begin{aligned} D_T(E) &= \sum_{\nu\in\mathbb{Z}} \frac{1}{\chi} d_\varepsilon(E)\,\exp\left[2\pi\mathrm{i}\nu\left(\frac{N_\varepsilon(E)}{\chi}+\frac{1}{2}\right)\right] \\ &= \sum_{\nu\in\mathbb{Z}} \frac{(-1)^\nu}{\chi} d_\varepsilon(E)\,\mathrm{e}^{2\pi\mathrm{i}\nu N_\varepsilon(E)/\chi}\,. \end{aligned} \tag{7.148}$$

The two-point correlation function of the approximate spectrum can now be expressed as, cf. (7.30),

$$\begin{aligned} R_{2,T}(s) &= \left\langle \frac{D_T(E'+\chi s/(2\bar{d}))\,D_T(E'-\chi s/(2\bar{d}))}{(\bar{d}/\chi)^2} \right\rangle_{E'} - 1 \\ &= \sum_{\nu,\nu'} \frac{(-1)^{\nu+\nu'}}{\bar{d}^2} \Big\langle d_\varepsilon\left(E'+\frac{\chi s}{2\bar{d}}\right) d_\varepsilon\left(E'-\frac{\chi s}{2\bar{d}}\right) \\ &\quad \times \exp\left[\frac{2\pi\mathrm{i}}{\chi}\left(\nu N_\varepsilon\left(E'+\frac{\chi s}{2\bar{d}}\right)+\nu' N_\varepsilon\left(E'-\frac{\chi s}{2\bar{d}}\right)\right)\right]\Big\rangle_{E'}\,. \end{aligned} \tag{7.149}$$

Upon splitting N_ε into $\overline{N}$ and $N_\varepsilon^{\mathrm{fl}}$ we see that in the semiclassical limit the $\overline{N}$-term causes rapid oscillations in E', since then the number of eigenvalues in the interval $I(E,\hbar)$ diverges. If we evaluated the E'-average by the method of stationary phase, we would obtain the condition

$$\nu\,\overline{d}\left(E'+\frac{\chi s}{2\overline{d}}\right)+\nu'\overline{d}\left(E'-\frac{\chi s}{2\overline{d}}\right)\stackrel{!}{=}0\,, \tag{7.150}$$

which for small $s/\overline{d}$ can only be fulfilled if $\nu'=-\nu$. Therefore, terms with $\nu'\neq-\nu$ are of relative order $\mathcal{O}(1/\overline{d})$ in comparison to those with $\nu'=-\nu$, and thus may be neglected in the semiclassical limit. One easily verifies that now the two-point correlation function can be written as

$$\begin{aligned} R_{2,T}(s) \sim & \underbrace{\left\langle \frac{d_\varepsilon(E'+\chi s/(2\overline{d}))\, d_\varepsilon(E'-\chi s/(2\overline{d}))}{\overline{d}}\right\rangle_{E'} - 1}_{=:R_{2,T}^{\text{non-osc}}(s)} \\ & \underbrace{+\frac{\partial^2}{\partial s\partial s'}\sum_{\nu\in\mathbb{Z}\setminus\{0\}}\frac{1}{(\pi\nu)^2}\left\langle \exp\left[\frac{2\pi\mathrm{i}\nu}{\chi}\Delta N_\varepsilon\right]\right\rangle_{E'}\Bigg|_{s'=s}}_{=:R_{2,T}^{\text{osc}}(s)}\,, \end{aligned} \tag{7.151}$$

$$\text{with}\quad \Delta N_\varepsilon := N_\varepsilon\left(E'+\frac{\chi s}{2\overline{d}}\right)-N_\varepsilon\left(E'-\frac{\chi s}{2\overline{d}}\right)\,.$$

The terms in the first line of (7.151) can be calculated as described in Sect. 7.3.1 and in diagonal approximation one finds

$$R_{2,T}^{\text{non-osc}}(s)\sim\frac{g}{T_H^2}\sum_\gamma \hat{\eta}(\varepsilon T_\gamma)\,\overline{\hat{\eta}(-\varepsilon T_\gamma)}\,|\mathcal{A}_\gamma|^2\exp\left(\frac{\mathrm{i}}{\hbar}\chi T_\gamma\frac{s}{\overline{d}}\right)\,. \tag{7.152}$$

This expression can be evaluated using a classical sum rule in a similar way as discussed in Sects. 7.3.2 and 7.5.1 yielding

$$R_{2,T}^{\text{non-osc}}(s)\sim-\frac{g}{2(\chi\pi s)^2}\,, \tag{7.153}$$

which is identical to the leading non-oscillating terms of the large s asymptotics (7.141)–(7.143) of $R_2^{\mathrm{GUE}}(s)$ $(g=1,\chi=1)$, $R_2^{\mathrm{GOE}}(s)$ $(g=2,\chi=1)$ and $R_2^{\mathrm{GSE}}(s)$ $(g=2,\chi=2)$, respectively.

Let us briefly demonstrate that this result is equivalent to the diagonal approximation of the form factor as determined in Sect. 7.3.2. If we evaluate the form factor on a test function ϕ, see (7.33), which for a moment shall be such that $\hat{\phi}$ is only supported at large s, we can formally write[5]

[5] Due to our choice of test functions we may ignore the singularity at $s=0$.

$$K(\tau) \sim \int_{\mathbb{R}} \frac{-g}{2(\chi\pi s)^2} \, \mathrm{e}^{2\pi \mathrm{i} \tau s} \, \mathrm{d}s \, . \tag{7.154}$$

By taking the second derivative with respect to τ this becomes

$$K''(\tau) \sim \int_{R} \frac{2g}{\chi^2} \, \mathrm{e}^{2\pi \mathrm{i} \tau s} \, \mathrm{d}s \sim \frac{2g}{\chi^2} \, \delta(\tau) \, . \tag{7.155}$$

Integrating the last equation from $-\tau$ to τ, $\tau > 0$ one obtains

$$K'(\tau) - K'(-\tau) \sim \frac{2g}{\chi^2} \quad \Leftrightarrow \quad K'(\tau) \sim \frac{g}{\chi^2} \, , \tag{7.156}$$

where we have used the symmetry of the form factor. By a further integration we find

$$K(\tau) \sim \frac{g}{\chi^2} \, \tau \tag{7.157}$$

which is identical to the diagonal approximation of the form factor as obtained in Sect. (7.3.2).

We can now turn our attention to the new terms $R_{2,T}^{\mathrm{osc}}(s)$ in (7.151). In the argument of the exponential we may again expand the (smooth) mean about E and split off the fluctuating part,

$$R_{2,T}^{\mathrm{osc}}(s) \sim \frac{\partial^2}{\partial s \partial s'} \sum_{\nu \in \mathbb{Z} \setminus \{0\}} \frac{\mathrm{e}^{\mathrm{i}\pi\nu(s+s')}}{(\pi\nu)^2} \left\langle \exp\left(\frac{2\pi \mathrm{i} \nu}{\chi} X(E') \right) \right\rangle_{E'} \Bigg|_{s'=s} \, , \tag{7.158}$$

$$\text{with} \quad X(E') := N_\varepsilon^{\mathrm{fl}}\left(E' + \frac{\chi s}{2\bar{d}}\right) - N_\varepsilon^{\mathrm{fl}}\left(E' - \frac{\chi s'}{2\bar{d}}\right) \, .$$

If we now assume that $X(E')$ is a Gaussian random variable (cf. our the discussion in the preceeding section) we have the identity

$$\left\langle \exp\left(\frac{2\pi \mathrm{i}\nu}{\chi} X(E') \right) \right\rangle_{E'} = \exp\left(-\frac{2\pi^2\nu^2}{\chi^2} \left\langle X^2(E') \right\rangle_{E'} \right) \, . \tag{7.159}$$

The variance can be calculated exactly as in Sect. (7.5.1) yielding

$$\left\langle X^2(E') \right\rangle_{E'} \sim \frac{g}{\pi^2} \log(s + s') \, . \tag{7.160}$$

Notice that present methods do not allow us to calculate the sub-leading term of this large-$s(s')$ expansion, cf. also the discussion in [55]. Since this variance has to be exponentiated we can only determine the s-dependence of $R_{2,T}^{\mathrm{osc}}(s)$ and not the normalisation. The leading terms in (7.158) derive from $|\nu| = 1$ with the derivatives taken in the first exponential, i.e.

$$R_{2,T}^{\mathrm{osc}}(s) \propto \frac{\cos(2\pi s)}{s^{2g/\chi^2}} \, . \tag{7.161}$$

This result is identical to the s-dependence of the leading oscillatory terms in (7.141)–(7.143). These in turn correspond to an expansion of the form factor about $\tau = 1$, where $K_{\mathrm{GUE}}(\tau)$ has a kink, $K_{\mathrm{GOE}}(\tau)$ has a kink in the second derivative and $K_{\mathrm{GSE}}(\tau)$ has a logarithmic singularity, see (7.36)–(7.38). Let us briefly demonstrate this connection for the last case by formally Fourier transforming (7.161) for $g = \chi = 2$ as in (7.154). In this case we have

$$K(\tau) = \int_{\mathbb{R}} \frac{\cos(2\pi s)}{4|s|} \, \mathrm{e}^{-2\pi \mathrm{i} \tau s} \, \mathrm{d}s \,, \tag{7.162}$$

where we have plugged in the correct pre-factor 1/4 from (7.143) and made use of the symmetry of $R_2(s)$. Thus, for the first derivative of $K(\tau)$ we find

$$\begin{aligned} K'(\tau) &= -\mathrm{i}\frac{\pi}{2} \int_{\mathbb{R}} \cos(2\pi s) \, \mathrm{sign}(s) \, \mathrm{e}^{-2\pi \mathrm{i} \tau s} \, \mathrm{d}s \\ &= -\mathrm{i}\frac{\pi}{4} \int_{\mathbb{R}} (2\theta(s) - 1) \left(\mathrm{e}^{-2\pi \mathrm{i} s(\tau-1)} + \mathrm{e}^{-2\pi \mathrm{i} s(\tau+1)} \right) \mathrm{d}s \\ &= -\frac{1}{4} \int_{\mathbb{R}} \delta(s) \left(\frac{\mathrm{e}^{-2\pi \mathrm{i} s(\tau-1)}}{\tau - 1} + \frac{\mathrm{e}^{-2\pi \mathrm{i} s(\tau+1)}}{\tau + 1} \right) \mathrm{d}s \\ &= -\frac{1}{4} \frac{2\tau}{\tau^2 - 1} \,, \end{aligned} \tag{7.163}$$

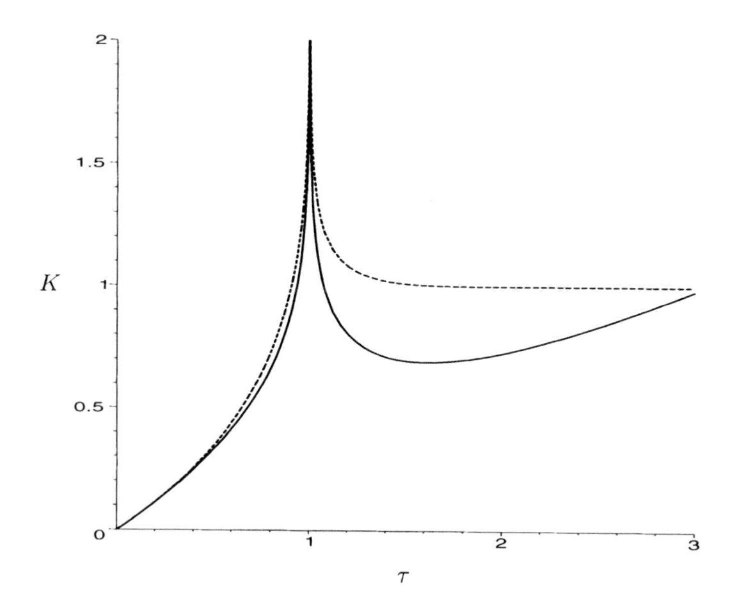

Fig. 7.7. We compare the sum of the diagonal form factor (7.80) and the bootstrap term (7.80) for time reversal invariant systems ($g = 2$) with half integer spin (*solid line*) to the GSE form factor (7.38) (*dashed line*), observing good agreement in the vicinities of $\tau = 0$ and $\tau = 1$.

where in the penultimate step we used a partial integration. Integrating again we find

$$K(\tau) = -\frac{1}{4}\log|\tau^2 - 1| = -\frac{1}{4}\log|\tau - 1| + \mathcal{O}(1) , \quad \tau \to 1 \tag{7.164}$$

where we have expanded about $\tau = 1$. This result is consistent with the behaviour of $K_{\mathrm{GSE}}(\tau)$ in the vicinity of $\tau = 1$ (7.38). We illustrate these findings in Fig. 7.7 where we plot the sum of the diagonal form factor (7.80) and the bootstrap-term (7.164) for generic time reversal invariant systems with half-integer spin together with the GSE form factor (7.38). As expected the two curves can hardly be distinguished near $\tau = 0$ and $\tau = 1$, whereas the behaviour for large τ, which is solely determined by the discreteness of the spectrum, is not reproduced by the semiclassics.

References

1. S.W. McDonald, A.N. Kaufman: Phys. Rev. Lett. **42**, 1189–1191 (1979)
2. M.V. Berry: Ann. Phys. (NY) **131**, 163–216 (1981)
3. O. Bohigas, M.J. Giannoni, C. Schmit: Phys. Rev. Lett. **52**, 1–4 (1984)
4. O. Bohigas: “Random matrix theory and chaotic dynamics”, in *Chaos and Quantum Physics*, ed. by M.J. Giannoni, A. Voros, J. Zinn Justin (1989), Les Houches, pp. 87–199
5. M.C. Gutzwiller: *Chaos in Classical and Quantum Mechanics* (Springer-Verlag, New York, 1990)
6. M.V. Berry, M. Tabor: Proc. R. Soc. London Ser. A **356**, 375–394 (1977)
7. M.L. Mehta: *Random Matrices*, 2nd edn. (Academic Press, San Diego, 1991)
8. T. Guhr, A. Müller-Groeling, H.A. Weidenmüller: Phys. Rep. **299**, 189–425 (1998)
9. A.M. Ozorio de Almeida: *Hamiltonian Systems: Chaos and Quantization* (Cambridge University Press, Cambridge, 1988)
10. M.V. Berry: “Some quantum-to-classical asymptotics”, in *Chaos and quantum physics*, ed. by M.J. Giannoni, A. Voros, J. Zinn Justin (1989), Les Houches, pp. 251–303
11. M. Tabor: *Chaos and Integrability in Nonlinear Dynamics* (John Wiley & Sons Inc., New York, 1989), an introduction
12. R. Blümel, W.P. Reinhardt: *Chaos in Atomic Physics* (Cambridge University Press, Cambridge, 1997)
13. J. Bolte: Open Sys. & Information Dyn. **6**, 167–226 (1999)
14. H.J. Stöckmann: *Quantum Chaos: An Introduction* (Cambridge University Press, Cambridge, 1999)
15. F. Haake: *Quantum Signatures of Chaos*, 2nd edn. (Springer-Verlag, Berlin Heidelberg, 2001)
16. J.H. Hannay, A.M. Ozorio de Almeida: J. Phys. A **17**, 3429–3440 (1984)
17. M.V. Berry: Proc. R. Soc. London Ser. A **400**, 229–251 (1985)
18. J.P. Keating: Nonlinearity **4**, 309–341 (1991)
19. E.B. Bogomolny, B. Georgeot, M.J. Giannoni, C. Schmit: Phys. Rev. Lett. **69**, 1477–1480 (1992)

20. J. Bolte, G. Steil, F. Steiner: Phys. Rev. Lett. **69**, 2188–2191 (1992)
21. J.P. Keating: Nonlinearity **4**, 277–307 (1991)
22. E.B. Bogomolny, B. Georgeot, M.J. Giannoni, C. Schmit: Phys. Rep. **291**, 219–324 (1997)
23. J. Bolte: Internat. J. Modern Phys. B **7**, 4451–4553 (1993)
24. E.B. Bogomolny, J.P. Keating: Phys. Rev. Lett. **77**, 1472–1475 (1996)
25. M. Sieber, K. Richter: Physica Scripta **T90**, 128–133 (2001)
26. M. Sieber: (2001), "Correlations between periodic orbits and their role in spectral statistics", talk given at the 7th Gentner Symposium on "Quantum Chaos", Ein Gedi, Israel (2001)
27. M. Sieber: J. Phys. A **35**, L613–L619 (2002)
28. S. Heusler: J. Phys. A **34**, L483–L490 (2001)
29. J. Bolte, S. Keppeler: J. Phys. A **32**, 8863–8880 (1999)
30. H.A. Kramers: Proc. Acad. Amst. **33**, 959–972 (1930)
31. E.P. Wigner: Nachrichten der Gesellschaft der Wissenschaften zu Göttingen, Mathematisch Physikalische Klasse pp. 546–559 (1932)
32. A. Messiah: *Quantum Mechanics. Vol. II*, Translated from the French by J. Potter (North-Holland Publishing Co., Amsterdam, 1962)
33. S. Keppeler, J. Marklof, F. Mezzadri: Nonlinearity **14**, 719–738 (2001)
34. R. Aurich, F. Steiner: Physica D **82**, 266–287 (1995)
35. E.B. Bogomolny, U. Gerland, C. Schmit: Eur. Phys. J. B **19**, 121–132 (2001)
36. M.C. Gutzwiller: J. Math. Phys. **12**, 343–358 (1971)
37. A. Shudo, Y. Shimizu, P. Šeba, J. Stein, H.J. Stöckmann, K. Życzkowski: Phys. Rev. E **49**, 3748–3756 (1994)
38. E.B. Bogomolny, U. Gerland, C. Schmit: Phys. Rev. E **59**, R1315–R1318 (1999)
39. K.F. Berggren, T. Ouchterlony: Found. Phys. **31**, 233–242 (2001)
40. A. Voros: Ann. Inst. H. Poincaré Sect. A (N.S.) **26**, 343–403 (1977)
41. R. Schubert: "Semiclassical localization in phase space", PhD thesis, Universität Ulm (2001)
42. M.V. Berry: Nonlinearity **1**, 399–407 (1988)
43. A. Bäcker, F. Steiner, P. Stifter: Phys. Rev. E **52**, 2463–2472 (1995)
44. R. Aurich, J. Bolte, F. Steiner: Phys. Rev. Lett. **73**, 1356–1359 (1994)
45. F. Steiner: "Quantum chaos", in *Schlaglichter der Forschung: zum 75. Jahrestag der Universität Hamburg 1994*, ed. by R. Ansorge (Dietrich Reimer Verlag, Berlin, Hamburg, 1994), pp. 543–564
46. E. Bogomolny, C. Schmit: Nonlinearity **6**, 523–547 (1993)
47. R. Aurich, A. Bäcker, F. Steiner: Int. J. Mod. Phys. B **11**, 805–849 (1997)
48. M.V. Berry, J.P. Keating: J. Phys. A **23**, 4839–4849 (1990)
49. R. Aurich, F. Steiner: Proc. R. Soc. London Ser. A **437**, 693–714 (1992)
50. M.V. Berry, J.P. Keating: Proc. R. Soc. London Ser. A **437**, 151–173 (1992)
51. R. Aurich, C. Matthies, M. Sieber, F. Steiner: Phys. Rev. Lett. **68**, 1629–1632 (1992)
52. R. Aurich, J. Bolte: Mod. Phys. Lett. **6**, 1691–1719 (1992)
53. J.P. Keating, M. Sieber: Proc. Roy. Soc. London Ser. A **447**, 413–437 (1994)
54. R. Aurich, F. Steiner: Phys. Rev. A **45**, 583–592 (1992)
55. S. Keppeler: J. Phys. A **33**, L503–L507 (2000)

Appendices

A The Poisson Summation Formula

The Poisson summation formula relates a sum over an L^1-function f evaluated at integer arguments to a sum over its Fourier transform (also required to be in L^1) by

$$\sum_{\boldsymbol{n}\in\mathbb{Z}^d} f(\boldsymbol{n}) = \sum_{\boldsymbol{\nu}\in\mathbb{Z}^d} \tilde{f}(\boldsymbol{\nu}) , \tag{A.1}$$

where

$$\tilde{f}(\boldsymbol{\xi}) := \int_{\mathbb{R}^d} f(\boldsymbol{x})\, \mathrm{e}^{2\pi \mathrm{i}\boldsymbol{x}\boldsymbol{\xi}}\, \mathrm{d}^d x . \tag{A.2}$$

Proof. Consider the function

$$F(\boldsymbol{x}) := \sum_{\boldsymbol{n}\in\mathbb{Z}^d} f(\boldsymbol{x}+\boldsymbol{n}) . \tag{A.3}$$

Obviously $F(\boldsymbol{x})$ is 1-periodic in all variables x_j and can thus be expanded into a Fourier series,

$$F(\boldsymbol{x}) = \sum_{\boldsymbol{m}\in\mathbb{Z}^d} c_{\boldsymbol{m}}\, \mathrm{e}^{2\pi \mathrm{i}\boldsymbol{m}\boldsymbol{x}} , \tag{A.4}$$

with

$$\begin{aligned} c_{\boldsymbol{m}} &= \int_0^1 \cdots \int_0^1 F(\boldsymbol{x})\, \mathrm{e}^{-2\pi \mathrm{i}\boldsymbol{m}\boldsymbol{x}}\, \mathrm{d}^d x \\ &= \sum_{\boldsymbol{n}\in\mathbb{Z}^d} \int_0^1 \cdots \int_0^1 f(\boldsymbol{x}+\boldsymbol{n})\, \mathrm{e}^{-2\pi \mathrm{i}\boldsymbol{m}\boldsymbol{x}}\, \mathrm{d}^d x . \end{aligned} \tag{A.5}$$

Changing variables to $\boldsymbol{y} = \boldsymbol{x}+\boldsymbol{n}$ and employing $\exp(-2\pi \mathrm{i}\boldsymbol{m}\boldsymbol{n}) = 1$ this simplifies to

$$\begin{aligned} c_{\boldsymbol{m}} &= \sum_{\boldsymbol{n}\in\mathbb{Z}^d} \int_{n_1}^{n_1+1} \cdots \int_{n_d}^{n_d+1} f(\boldsymbol{y})\, \mathrm{e}^{-2\pi \mathrm{i}\boldsymbol{m}\boldsymbol{y}}\, \mathrm{d}^d y \\ &= \int_{\mathbb{R}^d} f(\boldsymbol{y})\, \mathrm{e}^{-2\pi \mathrm{i}\boldsymbol{m}\boldsymbol{y}}\, \mathrm{d}^d y = \tilde{f}(\boldsymbol{m}) . \end{aligned} \tag{A.6}$$

Thus, we find

$$\sum_{\boldsymbol{n}\in\mathbb{Z}^d} f(\boldsymbol{x}+\boldsymbol{n}) = \sum_{\boldsymbol{m}\in\mathbb{Z}^d} \tilde{f}(\boldsymbol{m})\,\mathrm{e}^{2\pi\mathrm{i}\boldsymbol{m}\boldsymbol{x}}\;, \tag{A.7}$$

which, with $\boldsymbol{x}=0$, concludes the proof. A distributional version of (A.1) is given by

$$\sum_{\boldsymbol{n}\in\mathbb{Z}^d} \delta(\boldsymbol{x}-\boldsymbol{n}) = \sum_{\boldsymbol{\nu}\in\mathbb{Z}^d} \mathrm{e}^{2\pi\mathrm{i}\boldsymbol{x}\boldsymbol{\nu}}\;. \tag{A.8}$$

B Solution of the Scalar Transport Equation

We derive the solution of the scalar transport equation (i.e. the transport equation without spin) which is used in the Sects. 3.3, 3.7 and 3.8.

The transport equation is given by, cf. (3.40),

$$\frac{\mathrm{d}}{\mathrm{d}t}\,a_0(\boldsymbol{x},\boldsymbol{\xi},t) + \frac{1}{2}\nabla_{\boldsymbol{x}}\left[\nabla_{\boldsymbol{p}}H(\nabla_{\boldsymbol{x}}S,\boldsymbol{x})\right]\,a_0(\boldsymbol{x},\boldsymbol{\xi},t) = 0 \tag{B.1}$$

where S solves the Hamilton–Jacobi equation

$$H(\nabla_{\boldsymbol{x}}S,\boldsymbol{x}) + \frac{\partial S}{\partial t} = 0\;. \tag{B.2}$$

Whenever the arguments of the generating function are suppressed, one should read $S\equiv S(\boldsymbol{x},\boldsymbol{\xi},t)$. Similarly we drop the arguments of the Hamiltonian, i.e. in the following read $H\equiv H(\nabla_{\boldsymbol{x}}S,\boldsymbol{x})$. The first derivative in (B.1) was defined by (3.39),

$$\frac{\mathrm{d}}{\mathrm{d}t} := (\nabla_{\boldsymbol{p}}H)\nabla_{\boldsymbol{x}} + \frac{\partial}{\partial t}\;. \tag{B.3}$$

For a function F of variables $\{p_j\}_{j=1,\dots,d}$ and $\{x_j\}_{j=1,\dots,d}$ we write

$$\frac{\partial^2 F}{\partial p^2} \quad\text{and}\quad \frac{\partial^2 F}{\partial x\partial p} \tag{B.4}$$

for the $d\times d$ matrices whose elements are the second derivatives

$$\frac{\partial^2 F}{\partial p_j\partial p_k} \quad\text{and}\quad \frac{\partial^2 F}{\partial x_j\partial p_k}\;, \tag{B.5}$$

respectively. Thus, the transport equation (B.1) reads

$$\frac{\mathrm{d}}{\mathrm{d}t}\,a_0(\boldsymbol{x},\boldsymbol{\xi},t) + \frac{1}{2}\left[\mathrm{tr}\left(\frac{\partial^2 H}{\partial p^2}\frac{\partial^2 S}{\partial x^2} + \frac{\partial^2 H}{\partial p\partial x}\right)\right]\,a_0(\boldsymbol{x},\boldsymbol{\xi},t) = 0\;. \tag{B.6}$$

We will rewrite this equation with the help of the following relation.

Claim.

$$\operatorname{tr}\left(\frac{\partial^2 H}{\partial p^2}\frac{\partial^2 S}{\partial x^2}+\frac{\partial^2 H}{\partial p\partial x}\right)=-\frac{\mathrm{d}}{\mathrm{d}t}\log\det\frac{\partial^2 S}{\partial x\partial\xi}\ . \tag{B.7}$$

Proof.

$$\begin{aligned}-\frac{\mathrm{d}}{\mathrm{d}t}\log\det\frac{\partial^2 S}{\partial x\partial\xi}&=-\frac{\mathrm{d}}{\mathrm{d}t}\operatorname{tr}\log\frac{\partial^2 S}{\partial x\partial\xi}=\operatorname{tr}\left(-\frac{\mathrm{d}}{\mathrm{d}t}\log\frac{\partial^2 S}{\partial x\partial\xi}\right)\\&=\operatorname{tr}\left\{-\left[\frac{\mathrm{d}}{\mathrm{d}t}\left(\frac{\partial^2 S}{\partial x\partial\xi}\right)\right]\left(\frac{\partial^2 S}{\partial x\partial\xi}\right)^{-1}\right\}\ ,\end{aligned} \tag{B.8}$$

where in the last step we have taken advantage of cyclic permutations under the trace. For the expression in curly brackets we now find

$$\begin{aligned}&-\left[\frac{\mathrm{d}}{\mathrm{d}t}\left(\frac{\partial^2 S}{\partial x\partial\xi}\right)\right]\left(\frac{\partial^2 S}{\partial x\partial\xi}\right)^{-1}\\&\quad=-\left[(\nabla_{\boldsymbol{p}}H\cdot\nabla_{\boldsymbol{x}})\left(\frac{\partial^2 S}{\partial x\partial\xi}\right)+\frac{\partial^2}{\partial x\partial\xi}(-H)\right]\left(\frac{\partial^2 S}{\partial x\partial\xi}\right)^{-1}\\&\quad=\left[\nabla_{\boldsymbol{x}}\cdot\left(\nabla_{\boldsymbol{p}}H\frac{\partial^2 S}{\partial x\partial\xi}\right)-(\nabla_{\boldsymbol{p}}H\cdot\nabla_{\boldsymbol{x}})\left(\frac{\partial^2 S}{\partial x\partial\xi}\right)\right]\left(\frac{\partial^2 S}{\partial x\partial\xi}\right)^{-1}\\&\quad=\nabla_{\boldsymbol{x}}\cdot(\nabla_{\boldsymbol{p}}H)\\&\quad=\frac{\partial^2 H}{\partial p^2}\frac{\partial^2 S}{\partial x^2}+\frac{\partial^2 H}{\partial p\partial x}\ ,\end{aligned} \tag{B.9}$$

where in the first step we have used the Hamilton–Jacobi equation (B.2). Together with (B.7) this proves the claim.

With this result the transport equation simplifies to

$$\frac{\mathrm{d}}{\mathrm{d}t}a_0(\boldsymbol{x},\boldsymbol{\xi},t)-\frac{1}{2}\left[\frac{\mathrm{d}}{\mathrm{d}t}\log\det\left(\frac{\partial^2 S}{\partial x\partial\xi}(\boldsymbol{x},\xi,t)\right)\right]a_0(\boldsymbol{x},\boldsymbol{\xi},t)=0\ , \tag{B.10}$$

which is solved by

$$a_0(\boldsymbol{x},\boldsymbol{\xi},t)=\sqrt{\det\frac{\partial^2 S}{\partial x\partial\xi}(\boldsymbol{x},\boldsymbol{\xi},t)}\ . \tag{B.11}$$

Furthermore, from $S(\boldsymbol{x},\boldsymbol{\xi},0)=\boldsymbol{x}\boldsymbol{\xi}$, see (3.31), we conclude that this solution fulfills the initial condition $a_0(\boldsymbol{x},\boldsymbol{\xi},0)=1$ as required.

C Some Facts About the Groups SU(2) and SO(3)

We list some properties of the groups SU(2) and SO(3), their representations and the covering map $\varphi:\mathrm{SU}(2)\to\mathrm{SO}(3)$. Many of these relations can be found in the chapter on angular momentum of any standard textbook on quantum mechanics, like [1]; others may be looked up in textbooks on Lie groups and their representations, see e.g. [2, 3].

The Lie group SU(2) is the matrix group of unitary 2×2 matrices with determinant one. Its generators are the Pauli matrices

$$\sigma_1 \equiv \sigma_x = \begin{pmatrix} 0 & 1 \\ 1 & 0 \end{pmatrix} , \quad \sigma_2 \equiv \sigma_y = \begin{pmatrix} 0 & -\mathrm{i} \\ \mathrm{i} & 0 \end{pmatrix} \quad \text{and} \quad \sigma_3 \equiv \sigma_z = \begin{pmatrix} 1 & 0 \\ 0 & -1 \end{pmatrix} . \tag{C.1}$$

These give rise to the Lie Algebra $\mathfrak{su}(2)$,

$$[\sigma_j, \sigma_k] = \sigma_j \sigma_k - \sigma_k \sigma_j = 2\mathrm{i}\sigma_l \quad j,k,l \text{ cyclic} , \tag{C.2}$$

of angular momentum, where the Lie bracket is the matrix commutator. The Pauli matrices are hermitian and traceless,

$$\boldsymbol{\sigma}^\dagger = \boldsymbol{\sigma} , \quad \mathrm{tr}\, \sigma_j = 0 , \quad j = 1,2,3 , \tag{C.3}$$

and, simply viewed as matrices, one has the property

$$\sigma_j^2 = \mathbb{1}_2 , \quad j = 1,2,3 . \tag{C.4}$$

For $\boldsymbol{a}, \boldsymbol{b} \in \mathbb{R}^3$ one easily verifies by direct computation that

$$(\boldsymbol{\sigma a})(\boldsymbol{\sigma b}) = \boldsymbol{ab} + \mathrm{i}\boldsymbol{\sigma}(\boldsymbol{a} \times \boldsymbol{b}) , \tag{C.5}$$

which implies the useful formula

$$\begin{aligned} \mathrm{e}^{\mathrm{i}\boldsymbol{\sigma a}} &= \sum_{\nu=0}^{\infty} \frac{\mathrm{i}^\nu}{\nu!} (\boldsymbol{\sigma a})^\nu = \sum_{\nu=0}^{\infty} \frac{(-1)^\nu}{(2\nu)!} (\boldsymbol{\sigma a})^{2\nu} + \mathrm{i} \sum_{\nu=0}^{\infty} \frac{(-1)^\nu}{(2\nu+1)!} (\boldsymbol{\sigma a})^{2\nu+1} \\ &= \sum_{\nu=0}^{\infty} \frac{(-1)^\nu}{(2\nu)!} |\boldsymbol{a}|^{2\nu} + \mathrm{i} \sum_{\nu=0}^{\infty} \frac{(-1)^\nu}{(2\nu+1)!} |\boldsymbol{a}|^{2\nu} (\boldsymbol{\sigma a}) \\ &= \cos|\boldsymbol{a}| + \mathrm{i} \frac{\boldsymbol{\sigma a}}{|\boldsymbol{a}|} \sin|\boldsymbol{a}| . \end{aligned} \tag{C.6}$$

The unitary irreducible representations of SU(2) can be characterised by a number $s \in \mathbb{N}/2$ (physically the quantum number of angular momentum or spin, respectively). We denote by π_s the $2s+1$ dimensional irreducible representation of SU(2), which gives rise to a derived representation of $\mathfrak{su}(2)$ by

$$\mathrm{d}\pi_s(\sigma_j) := \frac{1}{\mathrm{i}} \frac{\mathrm{d}}{\mathrm{d}\lambda} \pi_s \left(\mathrm{e}^{\mathrm{i}\lambda\sigma_j} \right) . \tag{C.7}$$

Physically, $\mathrm{d}\pi_s(\boldsymbol{\sigma})$ is a normalised spin operator, cf. (3.5). For instance one finds

$$
\begin{aligned}
\mathrm{d}\pi_{1/2}(\boldsymbol{\sigma}) &= \boldsymbol{\sigma}\ , \\
\mathrm{d}\pi_1(\boldsymbol{\sigma}) &= \left(\begin{pmatrix} 0 & \sqrt{2} & 0 \\ \sqrt{2} & 0 & \sqrt{2} \\ 0 & \sqrt{2} & 0 \end{pmatrix}, \begin{pmatrix} 0 & -\mathrm{i}\sqrt{2} & 0 \\ \mathrm{i}\sqrt{2} & 0 & -\mathrm{i}\sqrt{2} \\ 0 & \mathrm{i}\sqrt{2} & 0 \end{pmatrix}, \begin{pmatrix} 2 & 0 & 0 \\ 0 & 0 & 0 \\ 0 & 0 & -2 \end{pmatrix} \right)^T , \\
\mathrm{d}\pi_{3/2}(\boldsymbol{\sigma}) &= \left(\begin{pmatrix} 0 & \sqrt{3} & 0 & 0 \\ \sqrt{3} & 0 & 2 & 0 \\ 0 & 2 & 0 & \sqrt{3} \\ 0 & 0 & \sqrt{3} & 0 \end{pmatrix}, \begin{pmatrix} 0 & -\mathrm{i}\sqrt{3} & 0 & 0 \\ \mathrm{i}\sqrt{3} & 0 & -2\mathrm{i} & 0 \\ 0 & 2\mathrm{i} & 0 & -\mathrm{i}\sqrt{3} \\ 0 & 0 & \mathrm{i}\sqrt{3} & 0 \end{pmatrix}, \begin{pmatrix} 3 & 0 & 0 & 0 \\ 0 & 1 & 0 & 0 \\ 0 & 0 & -1 & 0 \\ 0 & 0 & 0 & -3 \end{pmatrix} \right)^T
\end{aligned}
\tag{C.8}
$$

If we parameterise SU(2) as

$$
\mathrm{SU}(2) \ni g = \mathrm{e}^{\mathrm{i}\boldsymbol{\sigma}\boldsymbol{a}}\ , \quad \boldsymbol{a} \in \mathbb{R}^3\ , \quad |\boldsymbol{a}| \leq 2\pi\ , \tag{C.9}
$$

one easily verifies that the character $\chi_s(g) := \mathrm{tr}\,\pi_s(g)$ is given by

$$
\chi_s(g) = \sum_{m_s=-s}^{s} \mathrm{e}^{2\mathrm{i}m_s|\boldsymbol{a}|}\ . \tag{C.10}
$$

We will see below that $2|\boldsymbol{a}|$ has the meaning of a rotation angle, which we denote by

$$
\alpha := 2|\boldsymbol{a}|\ . \tag{C.11}
$$

Since this parameter completely determines the character, by an obvious abuse of notation, we also define

$$
\chi_s(\alpha) := \sum_{m_s=-s}^{s} \mathrm{e}^{\mathrm{i}m_s\alpha}\ . \tag{C.12}
$$

For instance one finds

$$
\begin{aligned}
\chi_{1/2}(\alpha) &= 2\cos(\alpha/2)\ , \\
\chi_1(\alpha) &= 1 + 2\cos\alpha\ , \\
\chi_{3/2}(\alpha) &= 2\cos(\alpha/2) + 2\cos(3\alpha/2)\ .
\end{aligned}
\tag{C.13}
$$

The Lie group SO(3) is the matrix group of orthogonal 3×3 matrices with determinant one, i.e. the group of proper rotations in three dimensional space. Its generators are given by

$$
\begin{aligned}
&E_1 \equiv E_x = \begin{pmatrix} 0 & 0 & 0 \\ 0 & 0 & -\mathrm{i} \\ 0 & \mathrm{i} & 0 \end{pmatrix}, \quad E_2 \equiv E_y = \begin{pmatrix} 0 & 0 & \mathrm{i} \\ 0 & 0 & 0 \\ -\mathrm{i} & 0 & 0 \end{pmatrix} \\
&\text{and} \quad E_3 \equiv E_z = \begin{pmatrix} 0 & -\mathrm{i} & 0 \\ \mathrm{i} & 0 & 0 \\ 0 & 0 & 0 \end{pmatrix},
\end{aligned}
\tag{C.14}
$$

which give rise to the Lie algebra $\mathfrak{so}(3)$,

$$[E_j, E_k] = \mathrm{i}E_l \; , \quad j,k,l \text{ cyclic} \; . \tag{C.15}$$

By rescaling $E_j \to 2E_j$ this becomes identical to (C.2), i.e. the groups SU(2) and SO(3) share the same Lie algebra.

Consider now the adjoint representation of SU(2), i.e. the representation of the group on its own algebra, defined by

$$\begin{aligned} \mathrm{Ad}_g : \; \mathfrak{su}(2) &\to \mathfrak{su}(2) \\ X &\mapsto gXg^{-1} \; . \end{aligned} \tag{C.16}$$

One easily verifies that this defines indeed a representation, i.e.

$$\mathrm{Ad}_{gh} = \mathrm{Ad}_g \circ \mathrm{Ad}_h \; . \tag{C.17}$$

Using the Pauli matrices as a basis in $\mathfrak{su}(2)$ we have $X = \boldsymbol{\sigma x}$, $\boldsymbol{x} \in \mathbb{R}^3$, and by

$$\mathrm{Ad}_g(\boldsymbol{\sigma x}) = g\,\boldsymbol{\sigma x}\,g^{-1} =: \boldsymbol{\sigma}\varphi(g)\boldsymbol{x} \tag{C.18}$$

the adjoint representation defines a map $\varphi : \mathrm{SU}(2) \to \mathrm{SO}(3)$. Let us explicitly determine this map for

$$g = \exp\left(-\mathrm{i}\frac{\alpha}{2}\boldsymbol{\sigma n}\right) \; , \quad \alpha \in [0, 4\pi) \; , \quad \boldsymbol{n} \in S^2 \subset \mathbb{R}^3 \; , \tag{C.19}$$

where we have expressed g in terms of α, see (C.11), and an axis $\boldsymbol{n}$. We obtain

$$\begin{aligned} gXg^{-1} &= (\cos(\alpha/2) - \mathrm{i}\boldsymbol{\sigma n}\sin(\alpha/2))(\boldsymbol{\sigma x})(\cos(\alpha/2) + \mathrm{i}\boldsymbol{\sigma n}\sin(\alpha/2)) \\ &= \cos^2(\alpha/2)\boldsymbol{\sigma x} + \mathrm{i}\cos(\alpha/2)\sin(\alpha/2)\left((\boldsymbol{\sigma x})(\boldsymbol{\sigma n}) - (\boldsymbol{\sigma n})(\boldsymbol{\sigma x})\right) \\ &\quad + \sin^2(\alpha/2)(\boldsymbol{\sigma n})(\boldsymbol{\sigma x})(\boldsymbol{\sigma n}) \; . \end{aligned} \tag{C.20}$$

With the relations

$$\begin{aligned} (\boldsymbol{\sigma x})(\boldsymbol{\sigma n}) - (\boldsymbol{\sigma n})(\boldsymbol{\sigma x}) &= 2\mathrm{i}\boldsymbol{\sigma}(\boldsymbol{x} \times \boldsymbol{n}) \\ \text{and} \quad (\boldsymbol{\sigma n})(\boldsymbol{\sigma x})(\boldsymbol{\sigma n}) &= (\boldsymbol{\sigma n})(\boldsymbol{xn} + \mathrm{i}\boldsymbol{\sigma}(\boldsymbol{x} \times \boldsymbol{n})) \\ &= -\boldsymbol{\sigma x} + 2(\boldsymbol{xn})(\boldsymbol{\sigma n}) \end{aligned} \tag{C.21}$$

this can be expressed as

$$\begin{aligned} gXg^{-1} &= \boldsymbol{\sigma x}\cos\alpha + \boldsymbol{\sigma}(\boldsymbol{n} \times \boldsymbol{x})\sin\alpha + (\boldsymbol{xn})(\boldsymbol{\sigma n})2\sin^2(\alpha/2) \\ &= \left[\boldsymbol{x}\cos\alpha + \boldsymbol{n} \times \boldsymbol{x}\sin\alpha + (\boldsymbol{xn})\boldsymbol{n}(1 - \cos\alpha)\right]\boldsymbol{\sigma} \\ &= \left[\left(\mathbb{1}_3\cos\alpha + A_{\boldsymbol{n}}\sin\alpha + \boldsymbol{n} \otimes \boldsymbol{n}(1 - \cos\alpha)\right)\boldsymbol{x}\right]\boldsymbol{\sigma} \; , \end{aligned} \tag{C.22}$$

with

$$A_{\boldsymbol{n}} = \begin{pmatrix} 0 & -n_3 & n_2 \\ n_3 & 0 & -n_1 \\ -n_2 & n_1 & 0 \end{pmatrix} = -\mathrm{i}\boldsymbol{E n} \; , \quad \boldsymbol{n} \otimes \boldsymbol{n} = \begin{pmatrix} n_1n_1 & n_1n_2 & n_1n_3 \\ n_2n_1 & n_2n_2 & n_2n_3 \\ n_3n_1 & n_3n_2 & n_3n_3 \end{pmatrix} \; . \tag{C.23}$$

Hence, we have found

$$\varphi\left(\mathrm{e}^{-\mathrm{i}\alpha\boldsymbol{\sigma n}/2}\right) = \mathbb{1}_3 \cos\alpha + A_{\boldsymbol{n}} \sin\alpha + \boldsymbol{n}\otimes\boldsymbol{n}(1-\cos\alpha) \;. \tag{C.24}$$

If we choose the coordinate system such that $\boldsymbol{n}\|\boldsymbol{e}_z$ we obtain

$$\begin{aligned}\varphi\left(\mathrm{e}^{-\mathrm{i}\alpha\boldsymbol{\sigma e}_z/2}\right) &= \mathbb{1}_3\cos\alpha + \begin{pmatrix}0 & -1 & 0\\ 1 & 0 & 0\\ 0 & 0 & 0\end{pmatrix}\sin\alpha + \begin{pmatrix}0&0&0\\0&0&0\\0&0&1\end{pmatrix}(1-\cos\alpha)\\ &= \begin{pmatrix}\cos\alpha & -\sin\alpha & 0\\ \sin\alpha & \cos\alpha & 0\\ 0 & 0 & 1\end{pmatrix} = \mathrm{e}^{-\mathrm{i}\alpha E_z}\;,\end{aligned} \tag{C.25}$$

i.e. $\varphi\left(\mathrm{e}^{-\mathrm{i}\alpha\boldsymbol{\sigma n}/2}\right)$ is a rotation about the z-axis by an angle α. In general one has

$$\varphi\left(\mathrm{e}^{-\mathrm{i}\alpha\boldsymbol{\sigma n}/2}\right) = \mathrm{e}^{-\mathrm{i}\alpha\boldsymbol{En}}\;, \tag{C.26}$$

a rotation about an axis $\boldsymbol{n}$ by an angle α. Thus, we see that $\varphi\left(\mathrm{e}^{-\mathrm{i}\alpha\boldsymbol{\sigma n}/2}\right)$ is indeed an element of SO(3), which we asserted before eq. (C.19). We have also shown our above claim that α is a rotation angle. One sees explicitly that the map φ provides a double covering of SO(3) since in (C.19) α runs from 0 to 4π.

D The Method of Stationary Phase

When doing semiclassical calculations one often encounters integrals of the form

$$I(\boldsymbol{y};\hbar) := \int_{\mathbb{R}^n} a(\boldsymbol{x},\boldsymbol{y})\,\mathrm{e}^{(\mathrm{i}/\hbar)\phi(\boldsymbol{x},\boldsymbol{y})}\,\mathrm{d}^n x\;,\quad \boldsymbol{x}\in\mathbb{R}^n\;,\;\boldsymbol{y}\in\mathbb{R}^m\;, \tag{D.1}$$

which can not be calculated explicitly. However, if one is only interested in the asymptotics for $\hbar\to 0$, one can employ the method of stationary phase. Here we only list the important formulae and give some motivation. For a complete proof the reader is referred to the literature, see e.g. [4, Chap. 2].

A simple reasoning concerning the asymptotics of (D.1) leads to the following picture. In the limit $\hbar\to 0$ the exponential oscillates rapidly, such that the contributions of all points with $\nabla_{\boldsymbol{x}}\phi\neq 0$ are negligible. This observation can be made into a precise statement in terms of the following lemma.

Lemma 2. *Let $\phi\in C^\infty(\mathbb{R}^n)$ be real valued, $a\in C_0^\infty(\mathbb{R}^n)$ and $\nabla_{\boldsymbol{x}}\phi\neq 0\;\forall\; x$ with $a(x)\neq 0$, then*

$$I(\hbar) := \int_{\mathbb{R}^n} a(\boldsymbol{x})\,\mathrm{e}^{(\mathrm{i}/\hbar)\phi(\boldsymbol{x})}\,\mathrm{d}^n x = \mathcal{O}(\hbar^N)\qquad \forall\, N>0\;,\;\hbar\to 0\;. \tag{D.2}$$

In order to proof the lemma one introduces the differential operator

$$L_x := \frac{-\mathrm{i}\hbar}{|\nabla_{\boldsymbol{x}}\phi|^2}(\nabla_{\boldsymbol{x}}\phi)\nabla_{\boldsymbol{x}} \,, \tag{D.3}$$

for which we have $L_x \mathrm{e}^{(\mathrm{i}/\hbar)\phi} = \mathrm{e}^{(\mathrm{i}/\hbar)\phi}$. Thus, one can insert an arbitrary power of L_x in front of the exponential, and by repeated partial integration one finds the required estimate. Lemma 2 can also be stated in the form $I(\hbar) = \mathcal{O}(\hbar^\infty)$.

For non-degenerate stationary points of ϕ, i.e. $a(\boldsymbol{x}_0) \neq 0$, $\nabla_{\boldsymbol{x}}\phi = 0$ and $\det\left(\partial^2\phi/(\partial x_k \partial x_l)\right) \neq 0$, one heuristically argues as follows. Expand ϕ about the stationary point x_0, terminate after the quadratic term and calculate the remaining Gaussian integral. It can be shown rigorously that this procedure indeed gives the leading term of the small-$\hbar$ asymptotics of $I(\boldsymbol{y}, \hbar)$, which in principle can be calculated to any order in $\hbar$. For the leading order we have the following precise statement:

Theorem 3. *Let $a \in C^\infty(\mathbb{R}^n)$ be such that $\{x \in \mathbb{R}^n \mid \exists\, y \in \mathbb{R}^m \text{ with } a(x,y) \neq 0\}$ is contained in a compact set in $\mathbb{R}^n$ and $\phi \in C^\infty(\mathbb{R}^n \times \mathbb{R}^m)$ real valued with non-degenerate stationary points $x_{0,j}(y)$. Then for (D.1) in leading order in $\hbar$ we have*

$$\begin{aligned} I(\boldsymbol{y};\hbar) = (2\pi\hbar)^{n/2} \sum_j a(\boldsymbol{x}_{0,j}(\boldsymbol{y}),\boldsymbol{y})\, \mathrm{e}^{(\mathrm{i}/\hbar)\phi(\boldsymbol{x}_{0,j}(\boldsymbol{y}),\boldsymbol{y})} \\ \times \frac{\exp\left(\mathrm{i}\frac{\pi}{4}\mathrm{sign}\left(\frac{\partial^2\phi}{\partial x_k \partial x_l}(\boldsymbol{x}_{0,j}(\boldsymbol{y}),\boldsymbol{y})\right)\right)}{\left|\det\left(\frac{\partial^2\phi}{\partial x_k \partial x_l}(\boldsymbol{x}_{0,j}(\boldsymbol{y}),\boldsymbol{y})\right)\right|^{1/2}} + \mathcal{O}(\hbar^{\frac{n}{2}+1}) \,. \end{aligned} \tag{D.4}$$

Here sign$A$ denotes the difference of the number of positive and negative eigenvalues of the matrix A. In the form of Theorem 3 the method of stationary phase is employed when deriving semiclassical trace formulae.

E Wigner-Weyl Calculus

We list some basic properties of the Wigner-Weyl calculus for multi-component wave equations and calculate the leading terms arising when one applies a Weyl operator to a WKB-like function.

We can associate with a differential operator $\hat{B}$ an object defined on classical phase space – its Weyl symbol $B(\boldsymbol{p},\boldsymbol{x})$ – by

$$(\hat{B}\Psi)(\boldsymbol{x}) = \frac{1}{(2\pi\hbar)^d} \int_{\mathbb{R}^d}\int_{\mathbb{R}^d} B\left(\boldsymbol{p}, \frac{\boldsymbol{x}+\boldsymbol{z}}{2}\right) \mathrm{e}^{(\mathrm{i}/\hbar)\boldsymbol{p}(\boldsymbol{x}-\boldsymbol{z})}\, \Psi(\boldsymbol{z})\, \mathrm{d}^d z\, \mathrm{d}^d p \,. \tag{E.1}$$

In general Ψ is multi-component, say $\Psi \in L^2(\mathbb{R}^d) \otimes \mathbb{C}^{2s+1}$, and thus $B(\boldsymbol{p},\boldsymbol{x})$ is matrix valued. Reverting the above reasoning, we can also associate an operator $\hat{B}$ with a more general symbol $B(\boldsymbol{p},\boldsymbol{x})$, which does not correspond to a differential operator, via (E.1). This procedure is known as Weyl quantisation and certain properties of symbols translate to properties of the operators, thus leading to so-called pseudo-differential operators, see e.g. [5] for an introduction. Examples of this correspondence, which appear in the present work, are the Dirac and Pauli operators

$$\begin{aligned}\hat{H}_{\mathrm{D}} &= c\boldsymbol{\alpha}\left(\frac{\hbar}{\mathrm{i}}\nabla - \frac{e}{c}\boldsymbol{A}(\boldsymbol{x})\right) + \beta mc^2 + e\phi(\boldsymbol{x}) ,\\ \hat{H}_{\mathrm{P}} &= -\frac{\hbar^2}{2m}\Delta - \frac{\hbar e}{mc}\boldsymbol{B}(\boldsymbol{x})\mathrm{d}\pi_s(\boldsymbol{\sigma})\end{aligned} \tag{E.2}$$

or the operator of angular momentum,

$$\hat{\boldsymbol{J}} = \hat{\boldsymbol{L}} + \frac{\hbar}{2}\mathrm{d}\pi_s(\boldsymbol{\sigma}) , \tag{E.3}$$

with the respective symbols

$$\begin{aligned}H_{\mathrm{D}}(\boldsymbol{p},\boldsymbol{x}) &= c\boldsymbol{\alpha}\left(\boldsymbol{p} - \frac{e}{c}\boldsymbol{A}(\boldsymbol{x})\right) + \beta mc^2 + e\phi(\boldsymbol{x}) ,\\ H_{\mathrm{P}}(\boldsymbol{p},\boldsymbol{x}) &= \frac{\boldsymbol{p}^2}{2m} - \frac{\hbar e}{mc}\boldsymbol{B}(\boldsymbol{x})\mathrm{d}\pi_s(\boldsymbol{\sigma}) \quad \text{and}\\ \boldsymbol{J}(\boldsymbol{p},\boldsymbol{x}) &= \boldsymbol{x}\times\boldsymbol{p} + \frac{\hbar}{2}\mathrm{d}\pi_s(\boldsymbol{\sigma}) .\end{aligned} \tag{E.4}$$

If an operator $\hat{B}$ can be represented by an integral kernel $K_B(\boldsymbol{x},\boldsymbol{y})$, i.e.

$$(\hat{B}\Psi)(\boldsymbol{x}) = \int_{\mathbb{R}^d} K_B(\boldsymbol{x},\boldsymbol{y})\,\Psi(\boldsymbol{y})\,\mathrm{d}^d y , \tag{E.5}$$

one obtains its Weyl symbol from

$$B(\boldsymbol{p},\boldsymbol{x}) = \int_{\mathbb{R}^d} K_B\left(\boldsymbol{x}+\frac{\boldsymbol{z}}{2},\boldsymbol{x}-\frac{\boldsymbol{z}}{2}\right)\mathrm{e}^{-(\mathrm{i}/\hbar)\boldsymbol{z}\boldsymbol{p}}\,\mathrm{d}^d z . \tag{E.6}$$

The inversion formula reads

$$K_B(\boldsymbol{x},\boldsymbol{y}) = \frac{1}{(2\pi\hbar)^d}\int_{\mathbb{R}^d} B\left(\boldsymbol{p},\frac{\boldsymbol{x}+\boldsymbol{y}}{2}\right)\mathrm{e}^{(\mathrm{i}/\hbar)\boldsymbol{p}(\boldsymbol{x}-\boldsymbol{y})}\,\mathrm{d}^d p . \tag{E.7}$$

The Weyl symbol of the projector $\hat{P}_\Psi$ projecting onto a state $\Psi \in L^2(\mathbb{R}^d)\otimes\mathbb{C}^{2s+1}$, i.e.

$$\left(\hat{P}_\Psi\,\Phi\right)(\boldsymbol{x}) = \Psi(\boldsymbol{x})\int_{\mathbb{R}^d}\Psi^\dagger(\boldsymbol{y})\,\Phi(\boldsymbol{y})\,\mathrm{d}^d y , \tag{E.8}$$

is known as the Wigner function

$$W_\Psi(\boldsymbol{p},\boldsymbol{x}) := \int_{\mathbb{R}^d}\Psi\left(\boldsymbol{x}+\frac{\boldsymbol{z}}{2}\right)\Psi^\dagger\left(\boldsymbol{x}-\frac{\boldsymbol{z}}{2}\right)\mathrm{e}^{-(\mathrm{i}/\hbar)\boldsymbol{z}\boldsymbol{p}}\,\mathrm{d}^d z , \tag{E.9}$$

one of the possible phase space representations of a quantum state. The expectation value of an observable $\hat{B}$ in a state Ψ can then be calculated in terms of a phase space integral over an expression involving the Weyl symbol $B(\boldsymbol{p},\boldsymbol{x})$ and the Wigner function $W_\Psi(\boldsymbol{p},\boldsymbol{x})$ as follows,

$$\int_{\mathbb{R}^d} \Psi^\dagger(\boldsymbol{x})\,\hat{B}\,\Psi(\boldsymbol{x})\,\mathrm{d}^d x = \frac{1}{(2\pi\hbar)^d}\int_{\mathbb{R}^d}\int_{\mathbb{R}^d} \mathrm{tr}\,(B(\boldsymbol{p},\boldsymbol{x})\,W_\Psi(\boldsymbol{p},\boldsymbol{x}))\,\mathrm{d}^d p\,\mathrm{d}^d x\,, \tag{E.10}$$

where tr denotes the matrix trace on $\mathbb{C}^{2s+1}$ as usual. We remark that from time to time we also find it convenient to use the following notation for the Weyl symbol of an observable $\hat{B}$,

$$W[\hat{B}](\boldsymbol{p},\boldsymbol{x}) \equiv B(\boldsymbol{p},\boldsymbol{x})\,. \tag{E.11}$$

An interesting formula of the Wigner-Weyl calculus is the Weyl symbol $C(\boldsymbol{p},\boldsymbol{x})$ of the commutator,

$$\hat{C} := [\hat{A},\hat{B}] \tag{E.12}$$

of two observables $\hat{A}$ and $\hat{B}$ with Weyl symbols

$$A(\boldsymbol{p},\boldsymbol{x}) = \sum_{k\geq 0}\hbar^k A_k(\boldsymbol{p},\boldsymbol{x}) \quad \text{and} \quad B(\boldsymbol{p},\boldsymbol{x}) = \sum_{k\geq 0}\hbar^k B_k(\boldsymbol{p},\boldsymbol{x})\,, \tag{E.13}$$

respectively. In such an expansion the leading order terms ($k=0$) are known as the principal symbols, whereas the next-to-leading order term ($k=1$) is the so-called sub-principal symbol. By a straightforward calculation one finds

$$\begin{aligned} C(\boldsymbol{p},\boldsymbol{x}) = {} & [A_0,B_0] + \hbar[A_0,B_1] + \hbar[B_0,A_1] + \frac{\hbar}{2\mathrm{i}}\{A_0,B_0\} - \frac{\hbar}{2\mathrm{i}}\{B_0,A_0\} \\ & + \hbar^2[A_0,B_2] + \hbar^2[A_2,B_0] + \hbar^2[A_1,B_1] \\ & + \frac{\hbar^2}{2\mathrm{i}}\{A_1,B_0\} - \frac{\hbar^2}{2\mathrm{i}}\{B_0,A_1\} + \frac{\hbar^2}{2\mathrm{i}}\{A_0,B_1\} - \frac{\hbar^2}{2\mathrm{i}}\{B_1,A_0\} \\ & - \frac{\hbar^2}{16}\{\{A_0,B_0\}\} - \frac{\hbar^2}{16}\{\{B_0,A_0\}\} + \mathcal{O}\left(\hbar^3\right)\,, \end{aligned} \tag{E.14}$$

where in the definition of the Poisson bracket,

$$\{A,B\} := \sum_{j=1}^{d}\left(\frac{\partial A}{\partial p_j}\frac{\partial B}{\partial x_j} - \frac{\partial A}{\partial x_j}\frac{\partial B}{\partial p_j}\right)\,, \tag{E.15}$$

the ordering is important, since the symbols are matrix valued. In particular, note that in general $\{B,B\} \neq 0$. The double Poisson bracket is defined as

$$\{\{A,B\}\} := \sum_{j,k=1}^{d}\left(\frac{\partial^2 A}{\partial p_j\partial p_k}\frac{\partial^2 B}{\partial x_j\partial x_k} - 2\frac{\partial^2 A}{\partial p_j\partial x_k}\frac{\partial^2 B}{\partial x_j\partial p_k} + \frac{\partial^2 A}{\partial x_j\partial x_k}\frac{\partial^2 B}{\partial p_j\partial p_k}\right). \tag{E.16}$$

For the important case that the principal symbols A_0 and B_0 are scalar, i.e. $A_0(\boldsymbol{p},\boldsymbol{x}) = a_0(\boldsymbol{p},\boldsymbol{x})\,\mathbb{1}_{2s+1}$ with a scalar function $a_0(\boldsymbol{p},\boldsymbol{x})$ and analogously for B_0, (E.14) reduces to

$$C(\boldsymbol{p},\boldsymbol{x}) = \frac{\hbar}{\mathrm{i}}\{A_0, B_0\} + \hbar^2[A_1, B_1] + \frac{\hbar^2}{\mathrm{i}}\{A_1, B_0\} + \frac{\hbar^2}{\mathrm{i}}\{A_0, B_1\} + \mathcal{O}\left(\hbar^3\right). \tag{E.17}$$

We now come to the main point of the appendix, the application of a Weyl operator to a rapidly oscillating (WKB) wave function

$$\Psi_{\mathrm{WKB}}(\boldsymbol{x}) = a_\hbar(\boldsymbol{x})\,\mathrm{e}^{\frac{\mathrm{i}}{\hbar}S(\boldsymbol{x})}\,, \quad a_\hbar(\boldsymbol{x}) = \sum_{k=0}^{\infty}\left(\frac{\hbar}{\mathrm{i}}\right)^k a_k(\boldsymbol{x})\,. \tag{E.18}$$

Whenever applying a Weyl operator $\hat{B}$, whose symbol has a semiclassical expansion of the form (E.13), to such an oscillating function the following theorem applies. The corresponding statement for homogeneous systems (i.e. without $\hbar$) can, e.g., be found in [6, Chap. 4.3].

Theorem 4. *Applying a Weyl operator $\hat{B}$ with a symbol $B(\boldsymbol{p},\boldsymbol{x})$ of the form (E.13) to a WKB wave function (E.18) yields in leading orders as $\hbar \to 0$,*

$$\begin{aligned}(\hat{B}\Psi_{\mathrm{WBK}})(\boldsymbol{x}) = \Bigg\{ & B_0(\nabla_{\boldsymbol{x}}S(\boldsymbol{x}),\boldsymbol{x})\,a_0(\boldsymbol{x}) + \frac{\hbar}{\mathrm{i}}\Bigg[B_0(\nabla_{\boldsymbol{x}}S(\boldsymbol{x}),\boldsymbol{x})\,a_1(\boldsymbol{x}) \\ & + \nabla_{\boldsymbol{p}}B_0(\nabla_{\boldsymbol{x}}S(\boldsymbol{x}),\boldsymbol{x})\,\nabla_{\boldsymbol{x}}a_0(\boldsymbol{x}) \\ & + \frac{1}{2}a_0(\boldsymbol{x})\nabla_{\boldsymbol{x}}[\nabla_{\boldsymbol{p}}B_0(\nabla_{\boldsymbol{x}}S(\boldsymbol{x}),\boldsymbol{x})] \\ & + B_1(\nabla_{\boldsymbol{x}}S(\boldsymbol{x}),\boldsymbol{x})\,a_0(\boldsymbol{x})\Bigg] + \mathcal{O}(\hbar^2)\Bigg\}\,\mathrm{e}^{\frac{\mathrm{i}}{\hbar}S(\boldsymbol{x})}\,.\end{aligned} \tag{E.19}$$

Proof. First expand the phase of the wave function about $\boldsymbol{z} = \boldsymbol{x}$:

$$\begin{aligned} S(\boldsymbol{z}) &= S(\boldsymbol{x}) + \nabla_{\boldsymbol{x}}S(\boldsymbol{x})(\boldsymbol{z}-\boldsymbol{x}) + R(\boldsymbol{z}) \\ \text{with}\quad R(\boldsymbol{z}) &:= S(\boldsymbol{z}) - S(\boldsymbol{x}) - \nabla_{\boldsymbol{x}}S(\boldsymbol{x})(\boldsymbol{z}-\boldsymbol{x})\,. \end{aligned} \tag{E.20}$$

Hence, we have the following properties for $R(\boldsymbol{z})$

$$R(\boldsymbol{x}) = 0\,, \quad \nabla_{\boldsymbol{z}}R(\boldsymbol{x}) = 0 \quad \text{and} \quad \frac{\partial^2 R(\boldsymbol{z})}{\partial z_j \partial z_k} = \frac{\partial^2 S(\boldsymbol{z})}{\partial z_j \partial z_k}\,. \tag{E.21}$$

Thus for the expression we are interested in we find,

$$
\begin{aligned}
&\mathrm{e}^{-(\mathrm{i}/\hbar)S(\boldsymbol{x})}\big(\hat{B}\mathrm{e}^{(\mathrm{i}/\hbar)S}a_\hbar\big)(\boldsymbol{x})\\
&= \frac{\mathrm{e}^{-(\mathrm{i}/\hbar)S(\boldsymbol{x})}}{(2\pi\hbar)^d}\int_{\mathbb{R}^d}\int_{\mathbb{R}^d} B\left(\boldsymbol{p},\frac{\boldsymbol{x}+\boldsymbol{z}}{2}\right)a_\hbar(\boldsymbol{z})\\
&\quad\times\exp\left(\frac{\mathrm{i}}{\hbar}[(\boldsymbol{p}-\nabla_{\boldsymbol{x}}S(\boldsymbol{x}))(\boldsymbol{x}-\boldsymbol{z})+S(\boldsymbol{x})+R(\boldsymbol{z})]\right)\mathrm{d}^dz\,\mathrm{d}^dp \qquad \text{(E.22)}\\
&= \frac{1}{(2\pi\hbar)^d}\int_{\mathbb{R}^d}\int_{\mathbb{R}^d} B\left(\boldsymbol{p},\frac{\boldsymbol{x}+\boldsymbol{z}}{2}\right)a_\hbar(\boldsymbol{z})\\
&\quad\times\exp\left(\frac{\mathrm{i}}{\hbar}(\boldsymbol{p}-\nabla_{\boldsymbol{x}}S(\boldsymbol{x}))(\boldsymbol{x}-\boldsymbol{z})+\frac{\mathrm{i}}{\hbar}R(\boldsymbol{z})\right)\mathrm{d}^dz\,\mathrm{d}^dp\,.
\end{aligned}
$$

Now expand $B\left(\boldsymbol{p},(\boldsymbol{x}+\boldsymbol{z})/2\right)\,\mathrm{e}^{(\mathrm{i}/\hbar)R(\boldsymbol{z})}a_\hbar(\boldsymbol{z})$ about $\boldsymbol{z}=\boldsymbol{x}$,

$$
\begin{aligned}
&B\left(\boldsymbol{p},\frac{\boldsymbol{x}+\boldsymbol{z}}{2}\right)\mathrm{e}^{(\mathrm{i}/\hbar)R(\boldsymbol{z})}a_\hbar(\boldsymbol{z})\\
&=\sum_{\nu=0}^{\infty}\frac{1}{\nu!}\left[(\boldsymbol{z}-\boldsymbol{x})\nabla_{\boldsymbol{z}}\right]^\nu\left(B\left(\boldsymbol{p},\frac{\boldsymbol{x}+\boldsymbol{z}}{2}\right)\mathrm{e}^{(\mathrm{i}/\hbar)R(\boldsymbol{z})}a_\hbar(\boldsymbol{z})\right)_{\boldsymbol{z}=\boldsymbol{x}}\,,
\end{aligned}
\qquad \text{(E.23)}
$$

where we only set $\boldsymbol{z}=\boldsymbol{x}$ after having taken all derivatives $\nabla_{\boldsymbol{z}}$. Using the relation

$$
(\boldsymbol{z}-\boldsymbol{x})\,\mathrm{e}^{(\mathrm{i}/\hbar)\boldsymbol{p}(\boldsymbol{x}-\boldsymbol{z})}=-\frac{\hbar}{\mathrm{i}}\nabla_{\boldsymbol{p}}\,\mathrm{e}^{(\mathrm{i}/\hbar)\boldsymbol{p}(\boldsymbol{x}-\boldsymbol{z})}\,, \qquad \text{(E.24)}
$$

after ν-fold partial integration one obtains

$$
\begin{aligned}
&\mathrm{e}^{-(\mathrm{i}/\hbar)S(\boldsymbol{x})}\big(\hat{B}\mathrm{e}^{(\mathrm{i}/\hbar)S}a_\hbar\big)(\boldsymbol{x})\\
&\quad=\frac{1}{(2\pi\hbar)^d}\int_{\mathbb{R}^d}\int_{\mathbb{R}^d}\mathrm{e}^{(\mathrm{i}/\hbar)(\boldsymbol{p}-\nabla_{\boldsymbol{x}}S(\boldsymbol{x}))(\boldsymbol{x}-\boldsymbol{z})}\\
&\quad\sum_{\nu=0}^{\infty}\frac{1}{\nu!}\left[\frac{\hbar}{\mathrm{i}}\nabla_{\boldsymbol{p}}\nabla_{\boldsymbol{z}}\right]^\nu\left(B\left(\boldsymbol{p},\frac{\boldsymbol{x}+\boldsymbol{z}}{2}\right)\mathrm{e}^{(\mathrm{i}/\hbar)R(\boldsymbol{z})}a_\hbar(\boldsymbol{z})\right)_{\boldsymbol{z}=\boldsymbol{x}}\mathrm{d}^dz\,\mathrm{d}^dp\,.
\end{aligned}
\qquad \text{(E.25)}
$$

Now the $\boldsymbol{z}$-integration can be done explicitly. At the same time we collect all terms up to order $\hbar^2$, whereby applying the properties (E.21). We find

$$
\begin{aligned}
&\mathrm{e}^{-(\mathrm{i}/\hbar)S(\boldsymbol{x})}(\hat{B}\mathrm{e}^{(\mathrm{i}/\hbar)S}a_\hbar)(\boldsymbol{x}) \\
&= \int_{\mathbb{R}^d} \delta\left(\boldsymbol{p} - \nabla_{\boldsymbol{x}} S(\boldsymbol{x})\right) \Bigg\{ B(\boldsymbol{p},\boldsymbol{x})\, a_0(\boldsymbol{x}) + \frac{\hbar}{\mathrm{i}} \Bigg[B(\boldsymbol{p},\boldsymbol{x})\, a_1(\boldsymbol{x}) \\
&\quad + \nabla_{\boldsymbol{p}} B(\boldsymbol{p},\boldsymbol{x})\, \nabla_{\boldsymbol{x}} a_0(\boldsymbol{x}) + \frac{1}{2} a_0(\boldsymbol{x})\, \nabla_{\boldsymbol{x}} \nabla_{\boldsymbol{p}} B(\boldsymbol{p},\boldsymbol{x}) \\
&\quad + \frac{1}{2} a_0(\boldsymbol{x}) \sum_{j,k=1}^{d} \frac{\partial^2 B}{\partial p_j \partial p_k}(\boldsymbol{p},\boldsymbol{x}) \frac{\partial^2 R}{\partial x_j \partial x_k}(\boldsymbol{x}) \Bigg] + \mathcal{O}(\hbar^2) \Bigg\} \mathrm{d}^d p \\
&= B(\nabla_{\boldsymbol{x}} S(\boldsymbol{x}),\boldsymbol{x})\, a_0(\boldsymbol{x}) + \frac{\hbar}{\mathrm{i}} \Bigg[B(\nabla_{\boldsymbol{x}} S(\boldsymbol{x}),\boldsymbol{x})\, a_1(\boldsymbol{x}) \\
&\quad + \nabla_{\boldsymbol{p}} B(\nabla_{\boldsymbol{x}} S(\boldsymbol{x}),\boldsymbol{x})\, \nabla_{\boldsymbol{x}} a_0(\boldsymbol{x}) + \frac{1}{2} a_0(\boldsymbol{x}) \nabla_{\boldsymbol{x}} [\nabla_{\boldsymbol{p}} B(\nabla_{\boldsymbol{x}} S(\boldsymbol{x}),\boldsymbol{x})] \Bigg] + \mathcal{O}(\hbar^2),
\end{aligned}
\tag{E.26}
$$

which proves the theorem. We remark that this result, certainly, holds analogously when applying a Weyl operator to a semiclassical ansatz for the time evolution kernel of the form

$$
K(\boldsymbol{x},\boldsymbol{y},t) = \int_{\mathbb{R}^d} a_\hbar(\boldsymbol{x},\boldsymbol{\xi},t)\, \mathrm{e}^{(\mathrm{i}/\hbar)(S(\boldsymbol{x},\boldsymbol{\xi},t) - \boldsymbol{y}\boldsymbol{\xi})}\, \mathrm{d}^d \xi \, . \tag{E.27}
$$

Then Theorem 4 provides the most general derivation of the leading order terms found, for example, in eqs. (3.34) or (3.192).

F Remarks on the Numerical Calculation of the Spectral Form Factor

When numerically calculating the spectral form factor $K(\tau)$ from a given set of unfolded eigenvalues $x_1, \ldots, x_N$ it is convenient to start from eq. (7.29) for the two-point correlation function which after Fourier transformation, cf. definition (7.32), reads ($N_I = N$)

$$
K_I(\tau) \sim \frac{1}{N} \sum_{n,n'=1}^{N} \mathrm{e}^{-2\pi \mathrm{i}\tau(x_n - x_{n'})} - \delta(\tau) = \underbrace{\frac{1}{N} \left| \sum_{n=1}^{N} \mathrm{e}^{-2\pi \mathrm{i}\tau x_n} \right|^2}_{=:F(\tau,N)} - \delta(\tau) \, . \tag{F.1}
$$

In the limit $N \to \infty$ the double sum $F(\tau,N)$ develops a δ-singularity at $\tau = 0$ which is canceled by the second term. The remaining terms then give rise to small-τ asymptotics like (7.39), (7.42) or (7.43). For finite N, however, the singularity of $F(\tau,N)$ at $\tau = 0$ is broadened which can be seen as follows. We expand the exponentials about $\tau = 0$ and resum the leading terms,

$$
\begin{aligned}
F(\tau,N) &= \frac{1}{N}\sum_{n,n'=1}^{N}\left[1-2\pi\mathrm{i}\tau(x_n-x_{n'})-2\pi^2\tau^2(x_n-x_{n'})^2+\mathcal{O}\left(\tau^3\right)\right]\\
&= \frac{1}{N}\sum_{n,n'=1}^{N}\left[1-2\pi^2\tau^2(x_n^2+x_{n'}^2-2x_nx_{n'})+\mathcal{O}\left(\tau^3\right)\right]\\
&= N\left[1-2\pi^2\tau^2(2\overline{x^2}-2\overline{x}^2)+\mathcal{O}\left(\tau^3\right)\right]\\
&= \underbrace{N\exp\left(-4\pi^2\tau^2(\overline{x^2}-\overline{x}^2)\right)}_{=:F_0(\tau,N)}+\mathcal{O}\left(\tau^3\right)\ ,
\end{aligned}
\tag{F.2}
$$

with

$$
\overline{x}=\frac{1}{N}\sum_{n=1}^{N}x_n \quad\text{and}\quad \overline{x^2}=\frac{1}{N}\sum_{n=1}^{N}x_n^2\ . \tag{F.3}
$$

Since the unfolded eigenvalues $\{x_n\}$ have a mean separation of one, their variance $\overline{x^2}-\overline{x}^2$ grows proportionally to N^2. Thus the broadened peak of $F(\tau,N)$ has a Gaussian shape and a width of order $1/N$.

We illustrate that $F_0(\tau,N)$ is indeed a good approximation to $F(\tau,N)$ for small values of τ in Fig. F.1 for a set of $N=400$ eigenvalues of a $\boldsymbol{\sigma p}$-rectangle. Therefore, we conclude that in numerical investigations any structure of $K(\tau)$ for $\tau\lesssim 1/N$ is hidden under the broadened peak $F_0(\tau,N)$.

An equivalent statement can be made on the basis of the oscillation which is clearly visible in Fig. F.1. The largest difference between unfolded eigenvalues x_n and $x_{n'}$, and thus the period of the fastest oscillation in eq. (F.1), is approximately given by N,

$$
\max_{1\le n,n'\le N}|x_n-x_{n'}|\approx N\ , \tag{F.4}
$$

since the unfolded eigenvalues have unit mean density. Thus, one cannot resolve any structure of $K(\tau)$ on scales below $1/N$. This determines the period of the oscillation visible in Fig. F.1 which is of the same order as the width of $F_0(\tau,N)$.

Since the size of the problematic region is of the order $1/N$ and thus rather small (even for $N=400$) these considerations should not substantially affect our results in the first place. However, in order to extract reasonable information from a numerically obtained form factor one has to average over an interval of width $\Delta\tau$, cf. Figs. 7.4 and 7.5 where we have chosen $\Delta\tau=0.2$. In the light of the previous discussion we thus cannot trust the data for $\tau\lesssim\Delta\tau/2=0.1$ as mentioned in Sect. 7.4, since it depends sensibly on how we treat the region with $\tau\lesssim 1/N$. We remark that this problem does not occur if instead of averaging over an interval $\Delta\tau$ one performs an ensemble average of the unsmoothed form factor over a certain set of similar systems. The problem is also absent from the study of quantised maps, where the values at which the form factor has to be evaluated are restricted to $\tau=\nu/N$, $\nu\in\mathbb{N}_0$, see e.g. [7].

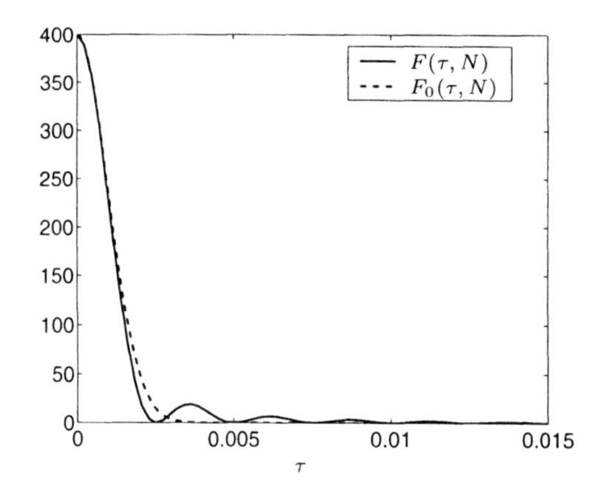

Fig. F.1. The double sum $F(\tau, N)$ and its approximation $F_0(\tau, N)$ for small τ are shown for a rectangular $\boldsymbol{\sigma p}$-billiard with spin $s = 1/2$ and coupling strength $\kappa = 32$.

References

1. A. Messiah: *Quantum Mechanics. Vol. II*, Translated from the French by J. Potter (North-Holland Publishing Co., Amsterdam, 1962)
2. A.O. Barut, R. Rączka: *Theory of group representations and applications* (PWN–Polish Scientific Publishers, Warsaw, 1977)
3. S. Sternberg: *Group theory and physics* (Cambridge University Press, Cambridge, UK, 1994)
4. A. Grigis, J. Sjöstrand: *Microlocal Analysis for Differential Operators*, London Mathematical Society Lecture Note Series (Cambridge University Press, Cambridge, 1994)
5. G.B. Folland: *Harmonic Analysis in Phase Space*, no. 122 in Annals of Mathematics Studies (Princeton University Press, Princeton, 1989)
6. J.J. Duistermaat: *Fourier Integral Operators* (Birkhäuser, Boston, 1996)
7. G. Haag, S. Keppeler: Nonlinearity **15**, 65–88 (2002)

Index

Springer Tracts in Modern Physics

154 **Applied RHEED**
Reflection High-Energy Electron Diffraction During Crystal Growth
By W. Braun 1999. 150 figs. IX, 222 pages

155 **High-Temperature-Superconductor Thin Films at Microwave Frequencies**
By M. Hein 1999. 134 figs. XIV, 395 pages

156 **Growth Processes and Surface Phase Equilibria in Molecular Beam Epitaxy**
By N.N. Ledentsov 1999. 17 figs. VIII, 84 pages

157 **Deposition of Diamond-Like Superhard Materials**
By W. Kulisch 1999. 60 figs. X, 191 pages

158 **Nonlinear Optics of Random Media**
Fractal Composites and Metal-Dielectric Films
By V.M. Shalaev 2000. 51 figs. XII, 158 pages

159 **Magnetic Dichroism in Core-Level Photoemission**
By K. Starke 2000. 64 figs. X, 136 pages

160 **Physics with Tau Leptons**
By A. Stahl 2000. 236 figs. VIII, 315 pages

161 **Semiclassical Theory of Mesoscopic Quantum Systems**
By K. Richter 2000. 50 figs. IX, 221 pages

162 **Electroweak Precision Tests at LEP**
By W. Hollik and G. Duckeck 2000. 60 figs. VIII, 161 pages

163 **Symmetries in Intermediate and High Energy Physics**
Ed. by A. Faessler, T.S. Kosmas, and G.K. Leontaris 2000. 96 figs. XVI, 316 pages

164 **Pattern Formation in Granular Materials**
By G.H. Ristow 2000. 83 figs. XIII, 161 pages

165 **Path Integral Quantization and Stochastic Quantization**
By M. Masujima 2000. 0 figs. XII, 282 pages

166 **Probing the Quantum Vacuum**
Pertubative Effective Action Approach in Quantum Electrodynamics and its Application
By W. Dittrich and H. Gies 2000. 16 figs. XI, 241 pages

167 **Photoelectric Properties and Applications of Low-Mobility Semiconductors**
By R. Könenkamp 2000. 57 figs. VIII, 100 pages

168 **Deep Inelastic Positron-Proton Scattering in the High-Momentum-Transfer Regime of HERA**
By U.F. Katz 2000. 96 figs. VIII, 237 pages

169 **Semiconductor Cavity Quantum Electrodynamics**
By Y. Yamamoto, T. Tassone, H. Cao 2000. 67 figs. VIII, 154 pages

170 **d–d Excitations in Transition-Metal Oxides**
A Spin-Polarized Electron Energy-Loss Spectroscopy (SPEELS) Study
By B. Fromme 2001. 53 figs. XII, 143 pages

171 **High-T_c Superconductors for Magnet and Energy Technology**
By B. R. Lehndorff 2001. 139 figs. XII, 209 pages

172 **Dissipative Quantum Chaos and Decoherence**
By D. Braun 2001. 22 figs. XI, 132 pages

173 **Quantum Information**
An Introduction to Basic Theoretical Concepts and Experiments
By G. Alber, T. Beth, M. Horodecki, P. Horodecki, R. Horodecki, M. Rötteler, H. Weinfurter, R. Werner, and A. Zeilinger 2001. 60 figs. XI, 216 pages

174 **Superconductor/Semiconductor Junctions**
By Thomas Schäpers 2001. 91 figs. IX, 145 pages

Springer Tracts in Modern Physics

Printing: Saladruck Berlin
Binding: Stürtz AG, Würzburg